solutions manual
for
students

solutions manual
for
students

to accompany

Paul A. Tipler

physics

for scientists and engineers

Fourth Edition

Volumes 2 & 3   Chapters 22–41

Frank J. Blatt
Professor Emeritus
Michigan State University

W. H. FREEMAN AND COMPANY
WORTH PUBLISHERS

Solutions Manual for Students, Volumes 2 & 3, Chapters 22–41
by Frank J. Blatt
to accompany
Tipler: *Physics for Scientists and Engineers,* Fourth Edition

Printed in the United States of America

ISBN: 1-57259-524-8

Printing: 2 3 4 5 — 03 02 01

W. H. Freeman and Company

41 Madison Avenue

New York. New York 10010

http://www.whfreeman.com

contents

CHAPTER **22**

The Electric Field I: Discrete Charge Distributions

1* • If the sign convention for charge were changed so that the charge on the electron were positive and the charge on the proton were negative, would Coulomb's law still be written the same?

Yes.

5* • How many coulombs of positive charge are there in 1 kg of carbon? Twelve grams of carbon contain Avogadro's number of atoms, with each atom having six protons and six electrons.

$Q = 6 \times n_C \times e$; $n_C = N_A(m_C/12)$ $Q = 6 \times 6.02 \times 10^{23} \times 10^3 \times 1.6 \times 10^{-19}/12$ C $= 4.82 \times 10^7$ C

9* •• Two uncharged conducting spheres with their conducting surfaces in contact are supported on a large wooden table by insulated stands. A positively charged rod is brought up close to the surface of one of the spheres on the side opposite its point of contact with the other sphere. (*a*) Describe the induced charges on the two conducting spheres, and sketch the charge distributions on them. (*b*) The two spheres are separated far apart and the charged rod is removed. Sketch the charge distributions on the separated spheres.

(*a*) On the sphere near the positively charged rod, the induced charge is negative and near the rod. On the other sphere, the net charge is positive and near the opposite side. This is shown in the diagram.

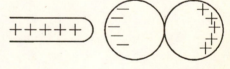

(*b*) When the spheres are separated and far apart and the rod has been removed, the induced charges are distributed uniformly over each sphere. The charge distributions are shown in the diagram.

13* •• Two equal charges of 3.0 μC are on the y axis, one at the origin and the other at $y = 6$ m. A third charge $q_3 = 2\ \mu$C is on the x axis at $x = 8$ m. Find the force on q_3.

Use Equ. 22-2 to find F_{13} and F_{23} $F_{13} = 8.43 \times 10^{-4}$ N i; $F_{23} = (5.39 \times 10^{-4}$ N$)(0.8\ i - 0.6\ j)$

$F_3 = F_{13} + F_{23}$ $F_3 = 1.27 \times 10^{-3}$ N $i - 3.24 \times 10^{-4}$ N j

---

17* •• A charge of $-1.0\ \mu$C is located at the origin, a second charge of 2.0 μC is located at $x = 0$, $y = 0.1$ m, and a third charge of 4.0 μC is located at $x = 0.2$ m, $y = 0$. Find the forces that act on each of the three charges.

Let $q_1 = -1.0\ \mu C$ at $(0, 0)$, $q_2 = 2\ \mu C$ at $(0, 0.1)$, and $q_3 = 4\ \mu C$ at $(0.2, 0)$.

1. Use Equ. 22-2 to find F_{21}, F_{31}, and F_{32}.

$F_{21} = 1.8\ N\,j$, $F_{31} = 0.899\ N\,i$,

$F_{32} = 0.643\ N\,j - 1.29\ N\,i$

2. $F_1 = F_{21} + F_{31}$; $F_2 = F_{12} + F_{32}$

$F_1 = 0.899\ N\,i + 1.8\ N\,j$; $F_2 = -1.29\ N\,i - 1.16\ N\,j$

3. $F_3 + F_1 + F_2 = 0$; $F_3 = -(F_1 + F_2)$

$F_3 = 0.391\ N\,i - 0.643\ N\,j$

21* •   A positive charge that is free to move but is at rest in an electric field E will

(a) accelerate in the direction perpendicular to E.

(b) remain at rest.

(c) accelerate in the direction opposite to E.

(d) accelerate in the same direction as E.

(e) do none of the above.

(d)

25* •   Two charges, each $+4\ \mu C$, are on the x axis, one at the origin and the other at $x = 8$ m. Find the electric field on the x axis at (a) $x = -2$ m, (b) $x = 2$ m, (c) $x = 6$ m, and (d) $x = 10$ m. (e) At what point on the x axis is the electric field zero? (f) Sketch E_x versus x.

(a) Use Equ. 22-7

$E = -8.99\times10^9 \times 4\times10^{-6}(1/2^2 + 1/10^2)\ N/C\ i$

$= -9.35\times10^3\ N/C\ i$

(b) Here the fields due to the two charges are oppositely directed

$E = 3.596\times10^4(1/2^2 - 1/6^2)\ N/C\ i = 7.99\times10^3\ N/C\ i$

(c) By symmetry, $E(6) = E(2)$

$E = -7.99\times10^3\ N/C\ i$

(d) By symmetry, $E(10) = E(-2)$

$E = 9.35\times10^3\ N/C\ i$

(e) Use symmetry argument

$E = 0$ at $x = 4$ m

(f) E_x versus x is shown

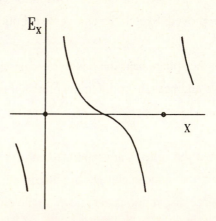

29* ••   Two equal positive charges of magnitude $q_1 = q_2 = 6.0$ nC are on the y axis at $y_1 = +3$ cm and $y_2 = -3$ cm. (a) What is the magnitude and direction of the electric field on the x axis at $x = 4$ cm? (b) What is the force exerted on a third charge $q_0 = 2$ nC when it is placed on the x axis at $x = 4$ cm?

(a) By symmetry, $E_y = 0$. Find E due to q_1 at $x = 4$ cm $E = kq_1/25\times10^{-4} = 2.158\times10^4\ N/C$

Total $E_x = 2E\times4/5$

$E_x = 3.45\times10^4\ N/C$; $E = 3.45\times10^4\ N/C\ i$

(b) $F = qE$ $F = 6.9 \times 10^{-5}$ N i

33* ••   A 5-μC point charge is located at $x = 1$ m, $y = 3$ m, and a -4-μC point charge is
located at $x = 2$ m, $y = -2$ m. (a) Find the magnitude and direction of the electric
field at $x = -3$ m, $y = 1$ m. (b) Find the magnitude and direction of the force on a
proton at $x = -3$ m, $y = 1$ m.

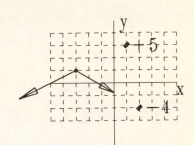

The diagram shows the electric field vectors at the point of interest due to the two
charges. Note that the E field due to the $+5$ μC charge makes an angle $\tan^{-1}(-0.5) =$
206.6° with the x axis; the E field due to the -4 μC charge makes an angle
$\tan^{-1}(-0.6) = -31°$ with the x axis

(a) 1. Find the magnitude of the two fields.
$E_5 = 5 \times 10^{-6} k / 20$
$E_{-4} = 4 \times 10^{-6} k / 34$

   2. Find the x and y components of the fields
$E_{x,5} = -0.224 \times 10^{-6} k$, $E_{x,-4} = 0.101 \times 10^{-6} k$
$E_{y,5} = -0.112 \times 10^{-6} k$; $E_{y,-5} = -0.061 \times 10^{-6} k$

   3. Find E
$E = -1.11 \times 10^3$ N/C i $- 1.55 \times 10^3$ N/C j
$E = 1.91 \times 10^3$ N/C at $\theta = 234.4°$

(b) $F = 1.6 \times 10^{-19} E$
$F = 3.06 \times 10^{-16}$ N at $\theta = 234.4°$

37* •   Which of the following statements about electric field lines is (are) *not* true?
(a) The number of lines leaving a positive charge or entering a negative charge is proportional to the charge.
(b) The lines begin on positive charges and end on negative charges.
(c) The density of lines (the number per unit area perpendicular to the lines) is proportional to the magnitude of
the field.
(d) Electric field lines cross midway between charges that have equal magnitude and sign.

(d)

41* •   Three equal positive point charges are situated at the corners of an equilateral triangle. Sketch the electric field
lines in the plane of the triangle.

A sketch of the field lines is shown.
Here we have assigned seven field lines
to each charge q.

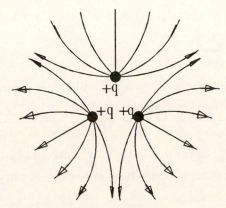

45* ••   An electron, starting from rest, is accelerated by a uniform electric field of 8×10^4 N/C that extends over a
distance of 5.0 cm. Find the speed of the electron after it leaves the region of uniform electric field.

$v^2 = 2ax; \ a = eE/m; \ v = \sqrt{2eEx/m}$ $v = \sqrt{2 \times 8 \times 10^4 \times 1.76 \times 10^{11} \times 0.05}$ m/s $= 3.75 \times 10^7$ m/s

---

49* ••   An electron starts at the position shown in Figure 22-33 with an initial speed $v_0 = 5 \times 10^6$ m/s at 45° to the x axis. The electric field is in the positive y direction and has a magnitude of 3.5×10^3 N/C. On which plate and at what location will the electron strike?

1. Note that $a = eE/m$ downward. Use Equ. 3-22 $x = (mv_0^2/eE)\sin 2\theta = 4.07$ cm on the lower plate

2. Find y_{max}; $y_{max} = m(v_0 \sin \theta)^2/2eE_y$ $y_{max} = 1.02$ cm; electron does not hit the upper plate.

53* ••   For a dipole oriented along the x axis, the electric field falls off as $1/x^3$ in the x direction and $1/y^3$ in the y direction. Use dimensional analysis to prove that, in any direction, the field far from the dipole falls off as $1/r^3$. Dimensionally, we can write $[E] = [kQ]/[L]^2$ and $[p] = [Q][L]$, where p represents the dipole. Thus the dimension of charge $[Q]$ is $[p]/[L]$, and the electric field has the dimension $[kp]/[L]^3$. This shows that the field E due to a dipole p falls off as $1/r^3$.

57* •••   A quadrupole consists of two dipoles that are close together, as shown in Figure 22-35. The effective charge at the origin is $-2q$ and the other charges on the y axis at $y = a$ and $y = -a$ are each $+q$. (a) Find the electric field at a point on the x axis far away so that $x \gg a$. (b) Find the electric field on the y axis far away so that $y \gg a$.

(a) We have, in effect, three charges: $+q$ at $(0, a)$, $+q$ at $(0, -a)$, and $-2q$ at $(0, 0)$. From the symmetry of the system it is evident that the field E along the x axis has no y component. The x component of E due to one of the charges $+q$ is

$$E_{+qx} = \frac{kq}{x^2 + a^2} \frac{x}{\sqrt{x^2 + a^2}} = \frac{kqx}{(x^2 + a^2)^{3/2}}. \text{ For the } -2q \text{ charge, } E_{-2qx} = \frac{-2kq}{x^2}. \text{ The total}$$

field along the x axis is $E_x = 2E_{+qx} + E_{-2qx}$. For $x \gg a$, $(x^2 + a^2)^{-3/2} \approx (1 - 3a^2/2x^2)/x^3$, and $E_x = -3kqa^2/x^4$.

(b) Along the y axis, $E_x = 0$ by symmetry. $E_y = kq/(y - a)^2 + kq/(y + a)^2 - 2kq/y^2$. Again using the binomial expansion one finds that for $y \gg a$, $E_y = 6kqa^2/y^4$.

61* ••   A molecule with electric dipole moment p is oriented so that p makes an angle θ with a uniform electric field E. The dipole is free to move in response to the force from the field. Describe the motion of the dipole. Suppose the electric field is nonuniform and is larger in the x direction. How will the motion be changed?

The dipole experiences a torque $\tau = pE \sin \theta$. In a uniform electric field, it will oscillate about its equilibrium orientation, $\theta = 0$. If the field is nonuniform and $dE/dx > 0$, the dipole will accelerate in the x direction as it oscillates about $\theta = 0$.

65* ••   A metal ball is positively charged. Is it possible for it to attract another positively charged ball? Explain.

Yes. A positively charged ball will induce a dipole on the metal ball, and if the two are in close proximity, the net force can be attractive.

69* ••   In copper, about one electron per atom is free to move about. A copper penny has a mass of 3 g. (a) What percentage of the free charge would have to be removed to give the penny a charge of 15 μC? (b) What would be the force of repulsion between two pennies carrying this charge if they were 25 cm apart? Assume that the pennies are point charges.

(a) Find the number of free electrons, $N = N_a$ From Example 22-1, $N = 2.84 \times 10^{22}$

Find n_e for a charge $q = -15 \ \mu$C $n_e = 15 \times 10^{-6}/e$

Fraction to be removed $= n_e/N$ $n_e/N = 15 \times 10^{-6}/(2.84 \times 10^{22} \times 1.6 \times 10^{-19}) = 3.3 \times 10^{-7}\%$

(b) Use Equ. 22-2

$$F = 225 \times 10^{-12} \times 8.99 \times 10^9 / 0.0625 \text{ N} = 32.4 \text{ N}$$

73* ••   A charge Q is located at $x = 0$ and a charge $4Q$ is at $x = 12.0$ cm. The force on a charge of -2 μC is zero if that charge is placed at $x = 4.0$ cm and is 126.4 N in the positive x direction if placed at $x = 8.0$ cm. Determine the charge Q.

1. Write F on -2 μC when at $x = 4$ cm $\qquad$ $126.4 \text{ N} = Q \times 2 \times 10^{-6} \times 8.99 \times 10^9 (4/16 \times 10^{-4} - 1/64 \times 10^{-4})$

2. Solve for Q $\qquad\qquad\qquad\qquad\qquad$ $Q = 3$ μC

77* ••   Two identical small spherical conductors (point charges), separated by 0.60 m, carry a total charge of 200 μC. They repel one another with a force of 120 N. (a) Find the charge on each sphere. (b) The two spheres are placed in electrical contact and then separated so that each carries 100 μC. Determine the force exerted by one sphere on the other when they are 0.60 m apart.

(a) Given: $q_1 + q_2 = 200$ μC; $kq_1q_2/0.36 = 120$ N $\qquad$ $q_2 = 2 \times 10^{-4} - q_1$; $8.99 \times 10^9 q_1 (2 \times 10^{-4} - q_1)/0.36 = 120$

$\qquad\qquad\qquad\qquad\qquad\qquad\qquad\qquad\qquad\qquad$ $q_1^2 - 2 \times 10^{-4} q_1 + 43.2 / 8.99 \times 10^9 = 0$

$\qquad$ Solve quadratic equation for q_1 $\qquad\qquad\qquad$ $q_1 = 28$ μC, $q_2 = 172$ μC; or $q_1 = 172$ μC, $q_2 = 28$ μC

(b) Now $q_1 = q_2 = 100$ μC; find F $\qquad\qquad$ $F = 8.99 \times 10^9 \times 10^{-8} / 0.36$ N $= 250$ N

81* ••   (a) Suppose that in Problem 80, $L = 1.5$ m, $m = 0.01$ kg, and $q = 0.75$ μC. What is the angle that each string makes with the vertical? (b) Find the angle that each string makes with the vertical if one mass carries a charge of 0.50 μC, the other a charge of 1.0 μC.

(a) 1. Use the expression given in Problem 80 $\qquad$ $\sin^2\theta \tan\theta = \dfrac{kq^2}{4L^2mg} = \dfrac{8.99 \times 10^9 \times (0.75 \times 10^{-6})^2}{4 \times 2.25 \times 0.01 \times 9.81}$

$\qquad$ 2. Since $\sin^2\theta \tan\theta << 1$, $\sin\theta \approx \tan\theta \approx \theta$ $\qquad$ $= 5.73 \times 10^{-3}$

$\qquad\qquad$ Solve for θ $\qquad\qquad\qquad\qquad\qquad\qquad\qquad$ $\theta^3 = 5.73 \times 10^{-3}$; $\theta = 0.179$ rad $= 10.25°$

(b) Repeat part (a) replacing q^2 by q_1q_2 $\qquad\qquad$ $\theta = 9.86°$

85* ••   An electron (charge $-e$, mass m) and a positron (charge $+e$, mass m) revolve around their common center of mass under the influence of their attractive coulomb force. Find the speed of each particle v in terms of e, m, k, and their separation r.

The force on each particle is ke^2/r^2. The centripetal acceleration of each particle is $v^2/(r/2)$. Using $F = ma$ one obtains $v = (ke^2/2mr)^{1/2}$.

89* •••   Repeat Problem 81 with the system located in a uniform electric field of 1.0×10^5 N/C that points vertically downward.

(a) Note that if the two charges are equal, each mass experiences an equal downward force of qE in addition to its weight mg. Thus, we may use the expression in Problem 80 provided we replace mg by $(mg + qE)$. As derived

in Problem 81, $\sin^2\theta \tan\theta = \dfrac{kq^2}{4L^2(mg + qE)} = \dfrac{8.99 \times 10^9 \times (0.75 \times 10^{-6})^2}{4 \times 2.25 (0.01 \times 9.81 + 0.075)} = 3.25 \times 10^{-3}$ and $\theta = 8.48°$.

(b) The downward forces on the two masses are not equal. Let the mass carrying the charge of 0.5 μC be m_1, and that carying the charge of 1.0 μC be m_2. Since we already know from part (a) that the angles are small, we shall make the small angle approximation $\sin\theta = \tan\theta = \theta$.

1. Write the horizontal and vertical forces on m_1 due to g, the charges q_1 and q_2, and tension T

$$F_{1x} = \frac{kq_1q_2}{L^2(\theta_1 + \theta_2)^2} = T_{1x}; \quad F_{1y} = m_1g + q_1E = T_{1y}$$

2. T_{2x} and T_{2y} are similar except for the subscripts

$$F_{2x} = \frac{kq_1q_2}{L^2(\theta_1 + \theta_2)^2} = T_{2x}; \quad F_{2y} = m_2g + q_2E = T_{2y}$$

3. $\theta_1 = T_{1x}/T_{1y}$; $\theta_2 = T_{2x}/T_{2y}$; find θ_1/θ_2

$$\theta_1/\theta_2 = (m_2g + q_2E)/(m_1g + q_1E)$$

4. Write the expression for $\theta_1 + \theta_2$

$$\theta_1 + \theta_2 = \frac{kq_1q_2}{L^2(\theta_1 + \theta_2)^2}\left(\frac{1}{m_1g + q_1E} + \frac{1}{m_2g + q_2E}\right)$$

5. Solve for $\theta_1 + \theta_2$

$$\theta_1 + \theta_2 = \left[\frac{kq_1q_2}{L^2}\left(\frac{1}{m_1g + q_1E} + \frac{1}{m_2g + q_2E}\right)\right]^{1/3}$$

6. Substitute numerical values for $m_1 = m_2 = m$ to determine $\theta_1 + \theta_2$, θ_1/θ_2, and θ_1 and θ_2

$$\theta_1 + \theta_2 = 0.287 \text{ rad} = 16.4°; \quad \theta_1/\theta_2 = 1.34$$
$$\theta_1 = 9.4°, \quad \theta_2 = 7.0°.$$

93* ••• A small bead of mass m, carrying a charge q, is constrained to slide along a thin rod of length L. Charges Q are fixed at each end of the rod (Figure 22-44). (*a*) Obtain an expression for the electric field due to the two charges Q as a function of x, where x is the distance from the midpoint of the rod. (*b*) Show that for $x \ll L$, the magnitude of the field is proportional to x. (*c*) Show that if q is of the same sign as Q, the force that acts on the object of mass m is always directed toward the center of the rod and is proportional to x. (*d*) Find the period of oscillation of the mass m if it is displaced by a small distance from the center of the rod and then released.

(*a*) Write the expression for E_x

$$E_x = kQ/(\tfrac{1}{2}L + x)^2 - kQ/(\tfrac{1}{2}L - x)^2$$

(*b*) For $x \ll L$, neglect x in denominator of (*a*)

$$E_x = -32kQx/L^3$$

(*c*) $F_x = qE_x$

$$F_x = -32kQqx/L^3; \text{ note that } F_x \text{ is proportional to } -x.$$

(*d*) $d^2x/dt^2 = -(16kQq/mL^3)x$; use Equs. 14-8, 14-12

$$T = (\pi/2)\sqrt{mL^3/2kQq}$$

CHAPTER 23

The Electric Field II: Continuous Charge Distributions

1* • A uniform line charge of linear charge density $\lambda = 3.5$ nC/m extends from $x = 0$ to $x = 5$ m. (*a*) What is the total charge? Find the electric field on the x axis at (*b*) $x = 6$ m, (*c*) $x = 9$ m, and (*d*) $x = 250$ m. (*e*) Find the field at $x = 250$ m, using the approximation that the charge is a point charge at the origin, and compare your result with that for the exact calculation in part (*d*).

(*a*) $Q = \lambda L$ $\qquad\qquad\qquad\qquad\qquad\qquad$ $Q = (3.5 \times 10^{-9} \times 5)$ C $= 17.5$ nC

(*b*), (*c*), (*d*) $E_x(x_0) = kQ/[x_0(x_0 - L)]$, Equ. 23-5 $\qquad$ $E_x(6) = 26.2$ N/C; $E_x(9) = 4.37$ N/C;
$\qquad\qquad\qquad\qquad\qquad\qquad\qquad\qquad\qquad\qquad\qquad$ $E_x(250) = 2.57 \times 10^{-3}$ N/C

(*e*) $E_x \approx kQ/x^2$ $\qquad\qquad\qquad\qquad\qquad\qquad$ $E_x(250) = 2.52 \times 10^{-3}$ N/C, within 2% of (*d*)

5* • For the disk charge of Problem 4, calculate exactly the electric field on the axis at distances of (*a*) 0.04 cm and (*b*) 5 m, and compare your results with those for parts (*b*) and (*c*) of Problem 4.

(*a*) Use Equ. 23-11; $r = 2.5$ cm, $\sigma = 3.6$ μC/m^2 $\qquad$ $E_x = 2.00 \times 10^5$ N/C

(*b*) Proceed as in (*a*) $\qquad\qquad\qquad\qquad\qquad\qquad$ $E_x = 2.54$ N/C

For $x = 0.04$ cm, the exact value of E_x is only 1.5% smaller than the approximate value obtained in the preceding problem. For $x = 5$ m, the exact and approximate values agree within less than 1%.

9* • Repeat Problem 8 for a disk of uniform surface charge density σ.

(*a*), (*b*), (*c*), (*d*), (*e*) Use Equ. 23-11; the results are $\qquad$ (*a*) $E_x = 0.804$ (*b*) $E_x = 0.553$ (*c*) $E_x = 0.427$
given in units of $\sigma/2\epsilon_0$. $\qquad\qquad\qquad\qquad\qquad$ (*d*) $E_x = 0.293$ (*e*) $E_x = 0.106$

(*f*) The field along the x axis is plotted in the adjoining figure. The x coordinates are in units of x/a and E is in units of $\sigma/2\epsilon_0$.

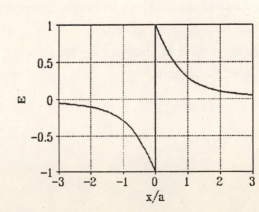

13★ ••   (a) A finite line charge of uniform linear charge density λ lies on the x axis from $x = 0$ to $x = a$. Show that the y component of the electric field at a point on the y axis is given by

$$E = \frac{k\lambda}{y} \sin \theta_1 = \frac{k\lambda}{y} \frac{a}{\sqrt{y^2 + a^2}}$$

where θ_1 is the angle subtended by the line charge at the field point. (b) Show that if the line charge extends from $x = -b$ to $x = a$, the y component of the electric field at a point on the y axis is given by

$$E_y = \frac{k\lambda}{y}(\sin \theta_1 + \sin \theta_2)$$

where $\sin \theta_2 = b/\sqrt{y^2 + b^2}$.

(a)   The line charge and the point $(0, y)$ are shown in the drawing. Also shown is the line element dx and the corresponding field dE. The y component of dE is then

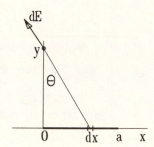

$$dE_y = \frac{k\lambda}{x^2 + y^2} \cos \theta \, dx = \frac{k\lambda y}{(x^2 + y^2)^{3/2}} dx.$$

Integrating dE_y from $x = 0$ to $x = a$ one obtains $E_y = \dfrac{k\lambda a}{y\sqrt{y^2 + a^2}} = \dfrac{k\lambda}{y} \sin \theta_1$, where here $\theta_1 = \theta$ shown in the drawing.

(b) Proceed as in part (a) but integrate dE_y from $x = -b$ to $x = a$. The result is

$$E_y = \frac{k\lambda}{y}\left(\frac{a}{\sqrt{y^2 + a^2}} + \frac{b}{\sqrt{y^2 + b^2}} \right) = \frac{k\lambda}{y}(\sin \theta_1 + \sin \theta_2), \text{ where } \theta_2 \text{ is the angle subtended by the line segment } b$$

at the point y.

17★ ••   True or false:

(a) Gauss's law holds only for symmetric charge distributions.

(b) The result that $E = 0$ inside a conductor can be derived from Gauss's law.

(a) False   (b) False

21★ •   A single point charge $q = +2 \, \mu C$ is at the origin. A spherical surface of radius 3.0 m has its center on the x axis at $x = 5$ m. (a) Sketch electric field lines for the point charge. Do any lines enter the spherical surface? (b) What is the net number of lines that cross the spherical surface, counting those that enter as negative? (c) What is the net flux of the electric field due to the point charge through the spherical surface?

(a) A sketch of the field lines and of the sphere is shown.

Three lines enter the sphere.

(b) The net number of lines crossing the surface is zero.

(c) The net flux is zero.

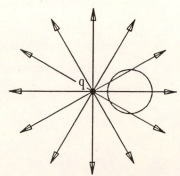

25* •   A point charge $q = +2\ \mu C$ is at the center of a sphere of radius 0.5 m. (a) Find the surface area of the sphere. (b) Find the magnitude of the electric field at points on the surface of the sphere. (c) What is the flux of the electric field due to the point charge through the surface of the sphere? (d) Would your answer to part (c) change if the point charge were moved so that it was inside the sphere but not at its center? (e) What is the net flux through a cube of side 1 m that encloses the sphere?

(a) $A = 4\pi r^2$ $A = 3.14\ \text{m}^2$

(b) Use Equ. 23-19 $E = 7.19 \times 10^4\ \text{N/C}$

(c) $\boldsymbol{E}\cdot\boldsymbol{n} = E$, so $\phi = EA$ $\phi = 2.26 \times 10^5\ \text{N}\cdot\text{m}^2/\text{C}$

(d) No change in ϕ if q is inside sphere

(e) Apply Gauss's law; ϕ unchanged $\phi = 2.26 \times 10^5\ \text{N}\cdot\text{m}^2/\text{C}$

29* ••   Explain why the electric field increases with r rather than decreasing as $1/r^2$ as one moves out from the center inside a spherical charge distribution of constant volume charge density.

The charge inside a sphere of radius r is proportional to r^3. The area of the sphere is proportional to r^2. Using Gauss's law one sees that the field must be proportional to $r^3/r^2 = r$.

33* ••   Consider two concentric conducting spheres (Figure 23-34). The outer sphere is hollow and initially has a charge $-7Q$ deposited on it. The inner sphere is solid and has a charge $+2Q$ on it. (a) How is the charge distributed on the outer sphere? That is, how much charge is on the outer surface and how much charge is on the inner surface? (b) Suppose a wire is connected between the inner and outer spheres. After electrostatic equilibrium is established, how much total charge is on the outside sphere? How much charge is on the outer surface of the outside sphere and how much is on the inner surface? Does the electric field at the surface of the inside sphere change when the wire is connected? If so, how? (c) Suppose we return to the original conditions in (a), with $+2Q$ on the inner sphere and $-7Q$ on the outer. We now connect the outer sphere to ground with a wire and then disconnect it. How much total charge will be on the outer sphere? How much charge will be on the inner surface of the outer sphere and how much will be on the outer surface?

(a) Since the outer sphere is conducting, the field in the thin shell must vanish. Therefore, $-2Q$, uniformly distributed, resides on the inner surface, and $-5Q$, uniformly distributed, resides on the outer surface.

(b) Now there is no charge on the inner surface, and $-5Q$ on the outer surface. The electric field just outside the surface of the inner sphere changes from a finite value to zero.

(c) In this case, the $-5Q$ is drained off, leaving no charge on the outer surface and $-2Q$ on the inner surface. The total charge on the outer sphere is then $-2Q$.

37* ••   Repeat Problem 35 for a sphere with volume charge density $\rho = C/r^2$ for $r < R$; $\rho = 0$ for $r > R$.

(a) The charge in a shell of thickness dr is $dq = 4\pi r^2 \rho dr = 4\pi C dr$. $Q = 4\pi C \int\limits_0^R dr = 4\pi CR$.

(b) For $r < R$, $Q_{in} = 4\pi Cr$ and by Gauss's law $E(r < R) = C/\epsilon_0 r$.

For $r > R$, $Q_{in} = 4\pi CR$ and $E(r > R) = CR/\epsilon_0 r^2$

A plot of $E(r)$ versus r/R is shown. Here $E(r)$ is in units of $C/\epsilon_0 R$.

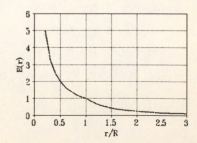

41* ••• A nonconducting solid sphere of radius a with its center at the origin has a spherical cavity of radius b with its center at the point $x = b$, $y = 0$ as shown in Figure 23-35. The sphere has a uniform volume charge density ρ. Show that the electric field in the cavity is uniform and is given by $E_y = 0$, $E_x = \rho b/3\epsilon_0$. (*Hint:* Replace the cavity with spheres of equal positive and negative charge densities.)

Using the *Hint* we shall find the x and y components of the field due to the uniform positive charge distribution of the solid sphere, and then the x and y components of the field due to a uniform negative charge distribution centered at $x = b$. We denote the field due to the solid positively charged sphere as E_+ and that due to the negatively charged sphere at $x = b$ by E_-. The field E_+ is $(4\pi/3)k\rho\, r$ and its x and y components are $E_{+x} = (4\pi/3)k\rho\, x$ and $E_{+y} = (4\pi/3)k\rho\, y$. For the negatively charged sphere, $E_- = -(4\pi/3)k\rho\, r'$, where r' is the radial distance from $x = b$, $y = 0$. We can again find the x and y components of E_- at a point (x, y); they are $E_{-x} = -(4\pi/3)k\rho\,(x - b)$ and $E_{-y} = -(4\pi/3)k\rho\, y$. Thus $E_x = E_{+x} + E_{-x} = (4\pi/3)k\rho b = \rho b/3\epsilon_0$ and $E_y = E_{+y} + E_{-y} = 0$. Since the system is symmetric for rotation about the x axis, $E_z = E_y = 0$.

45* •• A cylinder of length 200 m and radius 6 cm carries a uniform volume charge density of $\rho = 300$ nC/m^3. (*a*) What is the total charge of the cylinder? Use the formulas given in Problem 44 to calculate the electric field at a point equidistant from the ends at (*b*) $r = 2$ cm, (*c*) $r = 5.9$ cm, (*d*) $r = 6.1$ cm, and (*e*) $r = 10$ cm. Compare your results with those in Problem 43.

(*a*) $Q = \pi R^2 L \rho$	$Q = 679$ nC
(*b*), (*c*) Use Equ. 23-29*b*	$E_r(2 \text{ cm}) = 339$ N/C; $E_r(5.9 \text{ cm}) = 1.00$ kN/C
(*d*), (*e*) Use Equ. 23-29*a*	$E_r(6.1 \text{ cm}) = 1.00$ kN/C; $E_r(10 \text{ cm}) = 610$ N/C

49* •• Repeat Problem 44 with $\rho = C/r$.

(*a*) Find λ_{in} within a radius r for $r < R$
$$\lambda_{in} = \int_0^r 2\pi r(C/r)\,dr = 2\pi C r$$
Use Gauss's law to determine E
$$E = 2\pi C r / 2\pi\epsilon_0 r = C/\epsilon_0 \text{ for } r < R$$

(*b*) Find λ_{in} for $r > R$
$$\lambda_{in} = \int_0^R 2\pi r(C/r)\,dr = 2\pi C R$$
Find E using Gauss's law
$$E = CR/\epsilon_0 r$$

53* • An uncharged metal slab has square faces with 12-cm sides. It is placed in an external electric field that is perpendicular to its faces. The total charge induced on one of the faces is 1.2 nC. What is the magnitude of the electric field?

$\sigma = Q/L^2$; $E = Q/L^2\epsilon_0$ $E = 9.41$ kN/C

57* •• A positive point charge of magnitude 2.5 μC is at the center of an uncharged spherical conducting shell of inner radius 60 cm and outer radius 90 cm. (*a*) Find the charge densities on the inner and outer surfaces of the shell and the total charge on each surface. (*b*) Find the electric field everywhere. (*c*) Repeat (*a*) and (*b*) with a net charge of +3.5 μC placed on the shell.

(*a*) For 60 cm $< r <$ 90 cm, $E = 0$ $q_{in} = -2.5\ \mu$C; $\sigma_{in} = -2.5\times10^{-6}/4\pi\times0.6^2 = -0.553\ \mu$C/m^2

$q_{inner} + q_{outer} = 0$ $q_{out} = 2.5\ \mu$C; $\sigma_{out} = 2.5\times10^{-6}/4\pi\times0.9^2 = 0.246\ \mu$C/m^2

(*b*) For $r < 0.6$ m, $E = kq/r^2$ $E = 2.25\times10^4/r^2$ N/C

For 0.6 m $< r < 0.9$ m, conductor $E = 0$

For $r > 0.9$ m, $E = kq/r^2$ $\hspace{4cm}$ $E = 2.25 \times 10^4/r^2$ N/C

(c) Since $E = 0$ in the conductor, q_{inner} is again -2.5 μC and $\sigma_{inner} = -0.553$ μC/m^2. Now $q_{inner} + q_{outer} = 3.5$ μC; consequently, $q_{outer} = 6.0$ μC and $\sigma_{outer} = 0.59$ μC/m^2. The fields are $2.25 \times 10^4/r^2$ N/C for $r < 0.6$ m, zero within the shell, and $5.4 \times 10^4/r^2$ N/C for $r > 0.9$ m.

61* •   True or false:

(a) If there is no charge in a region of space, the electric field on a surface surrounding the region must be zero everywhere.

(b) The electric field inside a uniformly charged spherical shell is zero.

(c) In electrostatic equilibrium, the electric field inside a conductor is zero.

(d) If the net charge on a conductor is zero, the charge density must be zero at every point on the surface of the conductor.

(a) False (b) True (assuming there are no charges inside the shell) (c) True (d) False

---

65* ••   Suppose that the total charge on the conducting shell of Figure 23-37 is zero. It follows that the electric field for $r < R_1$ and $r > R_2$ points

(a) away from the center of the shell in both regions.

(b) toward the center of the shell in both regions.

(c) toward the center of the shell for $r < R_1$ and is zero for $r > R_2$.

(d) away from the center of the shell for $r < R_1$ and is zero for $r > R_2$.

(b)

69* ••   Equation 23-8 for the electric field on the perpendicular bisector of a finite line charge is different from Equation 23-9 for the electric field near an infinite line charge, yet Gauss's law would seem to give the same result for these two cases. Explain.

The two expressions agree if $r \ll L$, where L is the length of the line charge of finite length. For r of the same orde of magnitude as L or greater, the electric field does not have cylindrical symmetry and one cannot use Gauss's law to determine E.

73* ••   A nonuniform surface charge lies in the yz plane. At the origin, the surface charge density is $\sigma = 3.10$ μC/m^2. Other charged objects are present as well. Just to the right of the origin, the x component of the electric field is $E_x = 4.65 \times 10^5$ N/C. What is E_x just to the left of the origin?

Use Equ. 23-24 $\hspace{3cm}$ $E_{x,left} = E_{x,right} - \sigma/\epsilon_0 = 1.15 \times 10^5$ N/C

77* ••   An infinite plane in the xz plane carries a uniform surface charge density $\sigma_1 = 65$ nC/m^2. A second infinite plane carrying a uniform charge density $\sigma_2 = 45$ nC/m^2 intersects the xz plane at the z axis and makes an angle of $30°$ with the xz plane as shown in Figure 23-40. Find the electric field in the xy plane at (a) $x = 6$ m, $y = 2$ m and (b) $x = 6$ m, $y = 5$ m.

Find E_1 and E_2 for (6, 2) and (6, 5) $\hspace{2cm}$ $E_1(6, 2) = E_1(6, 5) = (\sigma_1/2\epsilon_0)\,j$;

$\hspace{7cm}$ $E_2(6, 2) = (\sigma_2/2\epsilon_0)(\sin 30° \, i - \cos 30° \, j)$;

$\hspace{7cm}$ $E_2(6, 5) = -E_2(6, 2)$

(a) $E(6, 2) = E_1(6, 2) + E_2(6, 2)$ $\hspace{2cm}$ $E(6, 2) = 1.27$ kN/C i + 1.47 kN/C j

(b) $E(6, 5) = E(6, 5) + E_2(6, 5)$ $\hspace{2cm}$ $E(6, 5) = -1.27$ kN/C i + 5.87 kN/C j

---

81* •• A rod of length L lies perpendicular to an infinitely long uniform line charge of charge density λ C/m (Figure 23-42). The near end of the rod is a distance d above the line charge. The rod carries a total charge Q uniformly distributed along its length. Find the force that the infinitely long line charge exerts on the rod.

Let y be the distance from the infinite line charge. The element of charge on the finite rod is $dq = (Q/L)dy$, and the field at the charge dq is $2k\lambda/y$. The force on the rod is

$$F = \frac{2k\lambda Q}{L}\int_{d}^{d+L}\frac{dy}{y} = \frac{2kQ\lambda}{L}\ln\left(\frac{L+d}{d}\right)$$

85* •• An infinite line charge λ is located along the z axis. A mass m that carries a charge q whose sign is opposite to that of λ is in a circular orbit in the xy plane about the line charge. Obtain an expression for the period of the orbit in terms of m, q, R, and λ, where R is the radius of the orbit.

For a circular orbit $mR\omega^2 = 2k\lambda q/R$; so $T = 1/f = 2\pi/\omega = 2\pi R\sqrt{\dfrac{m}{2kq\lambda}}$

89* •• A nonconducting cylinder of radius 1.2 m and length 2.0 m carries a charge of 50 μC uniformly distributed throughout the cylinder. Find the electric field *on the cylinder axis* at a distance of (*a*) 0.5 m, (*b*) 2.0 m, and (*c*) 20 m from the center of the cylinder.

We shall first solve in general terms and then insert appropriate numerical values. Let the origin of coordinates be at the center of the cylinder. Now consider a disk of radius R, the radius of the cylinder, and thickness dx. The charge carried by that disk is $dq = (Q/L)dx$, where Q is the total charge of the cylinder and L its length. The disk has an effective surface charge density $\sigma = Q/\pi R^2 L$. We can now use Equ. 23-11 to find the field due to this disk along its axis.

If the point of interest, P, is within the cylinder, the charge to the left of P will result in a field to the right, the charge to the right of P will give a field to the left. Thus,

$$E = 2\pi k\sigma\left[\int_{0}^{L/2+x}\left(1 - \frac{x}{\sqrt{x^2+R^2}}\right)dx - \int_{0}^{L/2-x}\left(1 - \frac{x}{\sqrt{x^2+R^2}}\right)dx\right].$$ Performing the indicated integrations one obtains

$E = 2\pi k\sigma[2x - \sqrt{(L/2+x)^2+R^2} + \sqrt{(L/2-x)^2+R^2}]$.

If P is beyond the end of the cylinder, the field at that point is given by

$$E = 2\pi k\sigma\int_{x-L/2}^{x+L/2}\left(1 - \frac{x}{\sqrt{x^2+R^2}}\right)dx = 2\pi k\sigma[L - \sqrt{(L/2+x)^2+R^2} + \sqrt{(L/2-x)^2+R^2}]$$ as before.

We can now substitute numerical values.

(*a*) For $x = 0.5$ m, $E = 119$ kN/C. (*b*) For $x = 2$ m, $E = 103$ kN/C. For $x = 20$ m, $E = 1.12$ kN/C; note that since the distance of 20 m is much greater than the length of the rod, we could have used $E_x \approx kQ/x^2 = 1.12$ kN/C.

93* ••• Two equal uniform line charges of length L lie on the x axis a distance d apart as shown in Figure 23-43. (*a*) What is the force that one line charge exerts on the other line charge? (*b*) Show that when $d \gg L$, the force tends toward the expected result of $k(\lambda L)^2/d^2$.

(*a*) Take $x = 0$ to be at the left hand end of the left rod. The the field at $x > L$ is $kQ/[x(x - L)]$. Now consider the right hand line charge. An element of charge in dx is $\lambda\, dx$ and experiences a force $E_x\lambda\, dx$. The total force due to the left hand line charge on the right hand line charge is therefore given by

$$F = k\lambda L \int_{L+d}^{2L+d} \frac{\lambda}{x(x-L)}\,dx = k\lambda^2 \ln\left[\frac{(d+L)^2}{d(2L+d)}\right]$$

(*b*) For $d \gg L$, the expression in the square brackets reduces to $1 + L^2/d^2$ to lowest order in L/d. We can now use the expansion $\ln(1+\epsilon) = \epsilon - \epsilon^2/2 + \cdots$ and again keeping only the first term obtain $F = k\lambda^2 L^2/d^2 = kQ^2/d^2$.

CHAPTER 24

Electric Potential

1* •   A uniform electric field of 2 kN/C is in the x direction. A positive point charge $Q = 3\ \mu$C is released from rest at the origin. (*a*) What is the potential difference $V(4\ \text{m}) - V(0)$? (*b*) What is the change in the potential energy of the charge from $x = 0$ to $x = 4$ m? (*c*) What is the kinetic energy of the charge when it is at $x = 4$ m? (*d*) Find the potential $V(x)$ if $V(x)$ is chosen to be (*d*) zero at $x = 0$, (*e*) 4 kV at $x = 0$, and (*f*) zero at $x = 1$ m.

(*a*) Use Equ. 24-2*b*; $\Delta V = -E\,\Delta x$ $\Delta V = -8$ kV

(*b*) $\Delta U = q\,\Delta V$ $\Delta U = -24$ mJ

(*c*) Use energy conservation $K = 24$ mJ

(*d*), (*e*), (*f*) Use Equ. 24-2*b*   (*d*) $V(x) = -(2\ \text{kV/m})x$; (*e*) $V(x) = 4\ \text{kV} - (2\ \text{kV/m})x$

                                             (*f*) $V(x) = 2\ \text{kV} - (2\ \text{kV/m})x$

5* •   A positive charge is released from rest in an electric field. Will it move toward a region of greater or smaller electric potential?

The positive charge will move toward lower potential energy, in this case toward the lower electric potential.

9* ••   Two identical masses m that carry equal charges q are separated by a distance d. Show that if both are released simultaneously their speeds when they are separated a great distance are $v/\sqrt{2}$, where v is the speed that one mass would have at a great distance from the other if it were released and the other held fixed.

If both are released, $2(\frac{1}{2}mv_b{}^2) = \Delta U$. If only one is released, $\frac{1}{2}mv^2 = \Delta U$. Hence $v_b = v/\sqrt{2}$.

13* •   Four 2-μC point charges are at the corners of a square of side 4 m. Find the potential at the center of the square (relative to zero potential at infinity) if (*a*) all the charges are positive, (*b*) three of the charges are positive and one is negative, and (*c*) two are positive and two are negative.

(*a*), (*b*), (*c*) Use Equ. 24-10; note that   (*a*) $V = (8.99\times10^9/2.83)(8\times10^{-6})$ V $= 25.4$ kV

$r_i = r = 4/\sqrt{2}$ m is the same for all charges.   (*b*) $V = 3.18\times10^9 \times 4\times10^{-6}$ V $= 12.7$ kV; (*c*) $V = 0$.

17* •   Two point charges q and q' are separated by a distance a. At a point $a/3$ from q and along the line joining the two charges the potential is zero. Find the ratio q/q'.

$q/(a/3) + q'/(2a/3) = 0$; $q/q' = -\frac{1}{2}$

21* ••   If E is known at just one point, can V be found at that point?

No

25* •   A point charge $q = 3.00 \ \mu C$ is at the origin. (a) Find the potential V on the x axis at $x = 3.00$ m and at $x = 3.01$ m. (b) Does the potential increase or decrease as x increases? Compute $-\Delta V/\Delta x$, where ΔV is the change in potential from $x = 3.00$ m to $x = 3.01$ m and $\Delta x = 0.01$ m. (c) Find the electric field at $x = 3.00$ m, and compare its magnitude with $-\Delta V/\Delta x$ found in part (b). (d) Find the potential (to three significant figures) at the point $x = 3.00$ m, $y = 0.01$ m, and compare your result with the potential on the x axis at $x = 3.00$ m. Discuss the significance of this result.

(a) Use Equ. 24-8 $V(3.00) = 8.99$ kV; $V(3.01) = 8.96$ kV

(b) From (a) V decreases with x $-\Delta V/\Delta x = 3.0$ kV/m

(c) $E = kq/r^2$ $E = 3.0$ kV/m, in agreement with (b)

(d) $V = kq/r = kq/(x^2 + y^2)^{1/2}$ $V(3.0, 0.01) = 8.99$ kV $= V(3.0, 0)$

For $y \ll x$, V is independent of y and the points $(x, 0)$ and (x, y) are at the same potential, i.e., on an equipotential surface.

29* •   The electric potential in some region of space is given by $V(x) = C_1 + C_2 x^2$, where V is in volts, x is in meters, and C_1 and C_2 are positive constants. Find the electric field E in this region. In what direction is E?

$E_x = -dV/dx = -2C_2 x$; with $C_2 > 0$, E points in the negative x direction.

33* •••   The electric potential in a region of space is given by $V = (2 \ V/m^2)x^2 + (1 \ V/m^2)yz$. Find the electric field at the point $x = 2$ m, $y = 1$ m, $z = 2$ m.

Use Equ. 24-18 $E_x = -4x$, $E_y = -z$, $E_z = -y$

Set $x = 2$ m, $y = 1$ m, $z = 2$ m $E = -8 \ V/m \ \boldsymbol{i} - 2 \ V/m \ \boldsymbol{j} - 1 \ V/m \ \boldsymbol{k}$

37* •   A charge of $q = +10^{-8}$ C is uniformly distributed on a spherical shell of radius 12 cm. (a) What is the magnitude of the electric field just outside and just inside the shell? (b) What is the magnitude of the electric potential just outside and just inside the shell? (c) What is the electric potential at the center of the shell? What is the electric field at that point?

(a) Use Gauss's law $E = 0$ for $r < 12$ cm; $E = kq/r^2$ for $r > 12$ cm. Just outside the shell $E = 6.24$ kV/m

(b) Use Equ. 24-23 $V = 749$ V just outside and inside the shell

(c) See Figure 24-12 for V; use Gauss's law for E $V(r = 0) = 749$ V; $E(r = 0) = 0$

41* ••   A rod of length L carries a charge Q uniformly distributed along its length. The rod lies along the y axis with its center at the origin. (a) Find the potential as a function of position along the x axis. (b) Show that the result obtained in (a) reduces to $V = kQ/x$ for $x \gg L$.

(a) The charge per unit length is $\lambda = Q/L$. Consider an element of length dy. The element of potential dV due to that line element is $dV = (k\lambda/r)dy$, where $r = \sqrt{x^2 + y^2}$. Then $V(x, 0)$ is obtained by integrating dV from $-L/2$ to $L/2$.

$$V(x,0) = \frac{kQ}{L} \int_{-\frac{L}{2}}^{\frac{L}{2}} \frac{dy}{\sqrt{x^2 + y^2}} = \frac{kQ}{L} \ln\left(\frac{\sqrt{x^2 + L^2/4} + L/2}{\sqrt{x^2 + L^2/4} - L/2} \right).$$

(b) Divide numerator and denominator within the parentheses by x and recall that $\ln(a/b) = \ln a - \ln b$.
Use the binomial expansion for $(1 + \epsilon)^{1/2} = 1 + \frac{1}{2}\epsilon - (1/8)\epsilon^2 + \cdots$ and $\ln(1 + \delta) = \delta - \delta^2/2 + \cdots$. Keeping only the lowest order terms in L/x one obtains $V = kQ/x$.

45* •• A disk of radius R carries a charge density $+\sigma_0$ for $r < a$ and an equal but opposite charge density $-\sigma_0$ for $a < r < R$. The total charge carried by the disk is zero. (a) Find the potential a distance x along the axis of the disk. (b) Obtain an approximate expression for $V(x)$ when $x >> R$.

(a) First, find the relation between a and R. Since the surface charge density is uniform, the magnitudes of the two charges are equal if $a^2 = R^2 - a^2$, or $a = R/\sqrt{2}$. The potential due to the positive charge is given by Equ. 24-20, where $a^2 = R^2/2$, i.e., $V_+ = 2\pi k\sigma_0[(x^2 + R^2/2)^{1/2} - x]$. To find V_- we integrate dV_-.

$$V_- = -2\pi\sigma_0 k \int_{R/\sqrt{2}}^{R} \frac{r\,dr}{\sqrt{x^2 + r^2}} = -2\pi\sigma_0 k\left(\sqrt{x^2 + R^2} - \sqrt{x^2 + R^2/2}\right). \quad V = 2\pi\sigma_0 k\left(2\sqrt{x^2 + R^2/2} - \sqrt{x^2 + R^2} - x\right).$$

(b) To determine V for $x >> R$, factor out x from the square roots and expand using the binomial expansion. Keeping only the lowest order terms, the expression in parentheses reduces to $R^4/16x^3$, and one obtains $V = \pi\sigma_0 kR^4/8x^3$ for $x >> R$.

49* •• Two very long, coaxial cylindrical shell conductors carry equal and opposite charges. The inner shell has radius a and charge $+q$; the other shell has radius b and charge $-q$. The length of each cylindrical shell is L. Find the potential difference between the shells.

For $a < r < b$, E_r is given by Equ. 23-9, i.e., $E_r = 2kq/Lr$. The potential difference $V_b - V_a$ is obtained by integration.

$$V_b - V_a = -\int_a^b (2kq/L)\frac{dr}{r} = -\frac{2kq}{L}\ln\left(\frac{b}{a}\right).$$

53* •• In Example 24-12 you derived the expression

$$V(r) = \frac{kQ}{2R}\left(3 - \frac{r^2}{R^2}\right)$$

for the potential inside a solid sphere of constant charge density by first finding the electric field. In this problem you derive the same expression by direct integration. Consider a sphere of radius R containing a charge Q uniformly distributed. You wish to find V at some point $r < R$. (a) Find the charge q' inside a sphere of radius r and the potential V_1 at r due to this part of the charge. (b) Find the potential dV_2 at r due to the charge in a shell of radius r' and thickness dr' at $r' > r$. (c) Integrate your expression in (b) from $r' = r$ to $r' = R$ to find V_2. (d) Find the total potential V at r from $V = V_1 + V_2$.

(a) Since the volume charge density is constant, $q' = Qr^3/R^3$. $V_1 = kq'/r = kQr^2/R^3$.

(b) $dV_2 = (k/r)dq'$, where now $dq' = 4\pi r'^2\rho\,dr' = (3Q/R^3)r'^2\,dr'$ since $\rho = 3Q/4\pi R^3$.

(c) Integrating dV_2 we have $V_2 = \frac{3kQ}{R^3}\int_r^R r'\,dr' = \frac{3kQ}{2R^3}(R^2 - r^2)$.

(d) $V = V_1 + V_2 = \frac{kQ}{2R}\left(3 - \frac{r^2}{R^2}\right)$.

57* •• Figure 24-24 shows a metal sphere carrying a charge $-Q$ and a point charge $+Q$. Sketch the electric field lines and equipotential surfaces in the vicinity of this charge system.

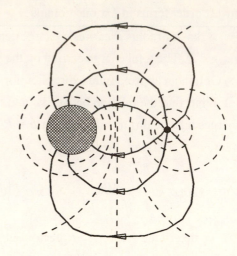

The electric field lines, shown as solid lines, and the equipotential surfaces (intersecting the plane of the paper), shown as dashed lines, are sketched in the adjacent figure. The point charge $+Q$ is the point at the right, and the metal sphere with charge $-Q$ is at the left. Near the two charges the equipotential surfaces are spheres, and the field lines are normal to the metal sphere at the sphere's surface.

61* • An infinite plane of charge has surface charge density 3.5 μC/m^2. How far apart are the equipotential surfaces whose potentials differ by 100 V?

$E = \sigma/2\epsilon_0 = -\Delta V/\Delta x; |\Delta x| = 2\epsilon_0 \Delta V/\sigma$ $\Delta x = 0.506$ mm

65* •• Charge is placed on two conducting spheres that are very far apart and connected by a long thin wire (Figure 24-25). The larger sphere has a diameter twice that of the smaller. Which sphere has the largest electric field near its surface? By what factor is it larger than that at the surface of the other sphere?

Both spheres are at the same potential so $V_L = V_S$. Hence, $kQ_L/R_L = kQ_S/R_S$. The fields near the surfaces are $E_S = kQ_S/R_S^2$, $E_L = kQ_L/R_L^2$. It follows that $E_S/E_L = R_L/R_S$ and that, therefore, the smaller sphere has the larger field near its surface, with $E_S = E_L(R_L/R_S)$.

69* • Two equal positive point charges $+Q$ are on the x axis. One is at $x = -a$ and the other is at $x = +a$. At the origin,
(a) $E = 0$ and $V = 0$.
(b) $E = 0$ and $V = 2kQ/a$.
(c) $E = (2kQ^2/a^2)\mathbf{i}$ and $V = 0$.
(d) $E = (2kQ^2/a^2)\mathbf{i}$ and $V = 2kQ/a$.
(e) none of the above is correct.
(b)

73* •• (a) V is constant on a conductor surface. Does this mean that σ is constant? (b) If E is constant on a conductor surface, does this mean that σ is constant? Does it mean that V is constant ?
(a) No (b) Yes; Yes

77* • If a conducting sphere is to be charged to a potential of 10,000 V, what is the smallest possible radius of the sphere such that the electric field will not exceed the dielectric strength of air?
For a sphere, $E_r = V(r)/r$. $r_{min} = V/E_{max}$ $r_{min} = 10^4/3 \times 10^6$ m $= 3.33$ mm

81* •• A Van de Graaff generator has a potential difference of 1.25 MV between the belt and the outer shell. Charge is supplied at the rate of 200 μC/s. What minimum power is needed to drive the moving belt?

$W = q\Delta V; \ P = dW/dt = \Delta V(dq/dt)$ $P = (200\times10^{-6} \times 1.25\times10^{6}) \ W = 250 \ W$

85* ••   A spherical conductor of radius R_1 is charged to 20 kV. When it is connected by a long, fine wire to a second conducting sphere far away, its potential drops to 12 kV. What is the radius of the second sphere?

1. Write the initial and final conditions $20 \ kV = k(Q_1 + Q_2)/R_1; \ 12 \ kV = kQ_1/R_1 = kQ_2/R_2$

2. Solve for R_1/R_2 $R_1/R_2 = 3/2; \ R_2 = (2/3)R_1$

89* ••   When you touch a friend after walking across a rug on a dry day, you typically draw a spark of about 2 mm. Estimate the potential difference between you and your friend before the spark.

E at breakdown is 3000 V/mm. So the potential difference is about 6000 V.

93* ••   A uniformly charged ring with a total charge of 100 μC and a radius of 0.1 m lies in the yz plane with its center at the origin. A meterstick has a point charge of 10 μC on the end marked 0 and a point charge of 20 μC on the end marked 100 cm. How much work does it take to bring the meterstick from a long distance away to a position along the x axis with the end marked 0 at $x = 0.2$ m and the other end at $x = 1.2$ m.

1. Use Equ. 24-20 to find $V(0.2 \ m)$ and $V(1.2 \ m)$ $V(0.2) = 4.02 \ MV; \ V(1.2) = 0.747 \ MV$

2. $W = q_{1.2}V(1.2) + q_{0.2}V(0.2)$ $W = 14.9 \ J + 40.2 \ J = 55.1 \ J$

97* •••   Three concentric conducting spherical shells have radii a, b, and c such that $a < b < c$. Initially, the inner shell is uncharged, the middle shell has a positive charge Q, and the outer shell has a negative charge $-Q$. (*a*) Find the electric potential of the three shells. (*b*) If the inner and outer shells are now connected by a wire that is insulated as it passes through the middle shell, what is the electric potential of each of the three shells, and what is the final charge on each shell?

(*a*) Since the total charge is zero, $V(r \geq c) = 0$, so $V(c) = 0$. Between the outer and middle shells the field is $E_r = kQ/r^2$, so the potential difference between c and b is $kQ(1/b - 1/c)$, and since $V(c) = 0$, $V(b) = kQ(1/b - 1/c)$. The inner shell carries no charge, so the field between $r = a$ and $r = b$ is zero and $V(a) = V(b)$.

(*b*) When the inner and outer shell are connected their potentials are equal. Also $Q_a + Q_c = -Q$, the initial charge on the outer shell. As before, $V(c) = 0 = V(a)$. In the region between the $r = a$ and $r = b$, the field is kQ_a/r^2 and the potential at $r = b$ is then $V(b) = kQ_a(1/b - 1/a)$. The enclosed charge for $b < r < c$ is $Q_a + Q$, and by Gauss's law the field in this region is $k(Q_a + Q)/r^2$ and the potential difference between b and c is

$V(c) - V(b) = k(Q_a + Q)(1/c - 1/b) = -V(b)$ since $V(c) = 0$.

We now have two expressions for $V(b)$ which can be used to determine Q_a. One obtains

$Q_a = -Q\dfrac{a(c - b)}{b(c - a)}$ and $Q_c = -Q\dfrac{c(b - a)}{b(c - a)}$ and $V(b) = kQ\dfrac{(c - b)(b - a)}{b^2(c - a)}$.

CHAPTER 25

Electrostatic Energy and Capacitance

1* • Three point charges are on the x axis: q_1 at the origin, q_2 at $x = 3$ m, and q_3 at $x = 6$ m. Find the electrostatic potential energy for (a) $q_1 = q_2 = q_3 = 2\ \mu C$, (b) $q_1 = q_2 = 2\ \mu C$ and $q_3 = -2\ \mu C$, and (c) $q_1 = q_3 = 2\ \mu C$ and $q_2 = -2\ \mu C$.

1. List $r_{1,2}, r_{1,3}$, and $r_{2,3}$ $r_{1,2} = 3$ m, $r_{1,3} = 6$ m, $r_{2,3} = 3$ m

(a), (b), (c) Use Equ. 25-1   (a) $U = 8.99 \times 10^9 \times 4 \times 10^{-12}(5/6)$ J $= 30.0$ mJ

                                   (b) $U = -5.99$ mJ (c) $U = -18.0$ mJ

5* •• Four charges are at the corners of a square centered at the origin as follows: q at $(-a, +a)$; $2q$ at (a, a); $-3q$ at $(a, -a)$; and $6q$ at $(-a, -a)$. A fifth charge $+q$ is placed at the origin and released from rest. Find its speed when it is a great distance from the origin.

1. Find $V(0)$ $V(0) = (kq/a\sqrt{2})(1 + 2 - 3 + 6) = 6kq/a\sqrt{2}$

2. $\frac{1}{2}mv^2 = qV(0)$ $v = q\sqrt{6\sqrt{2}k/ma}$

9* • An isolated spherical conductor of radius 10 cm is charged to 2 kV. (a) How much charge is on the conductor? (b) What is the capacitance of the sphere? (c) How does the capacitance change if the sphere is charged to 6 kV?

(a) $Q = Vr/k$ $Q = 2 \times 10^3 \times 0.1/8.99 \times 10^9$ C $= 22.2$ nC

(b) $C = Q/V = r/k$ $C = 11.1$ pF

(c) No change of C with V

13* •• If the potential difference of a parallel-plate capacitor is doubled by changing the plate separation without changing the charge, by what factor does its stored electric energy change?

$U = \frac{1}{2}QV$ Doubling V doubles U

17* •• (a) A 3-μF capacitor is charged to 100 V. How much energy is stored in the capacitor? (b) How much additional energy is required to charge the capacitor from 100 to 200 V?

(a) $U = \frac{1}{2}CV^2$ $U = 15$ mJ

(b) $U_b = 4U_a$ $\Delta U = 45$ mJ

21* •   A parallel-plate capacitor with a plate area of 2 m² and a separation of 1.0 mm is charged to 100 V. (*a*) What is the electric field between the plates? (*b*) What is the energy per unit volume in the space between the plates? (*c*) Find the total energy by multiplying your answer to part (*b*) by the total volume between the plates. (*d*) Find the capacitance *C*. (*e*) Calculate the total energy from $U = \frac{1}{2}CV^2$, and compare your answer with your result for part (*c*).

(*a*) $E = V/d$ $E = 100/10^{-3}$ V/m = 100 kV/m

(*b*) $u = \epsilon_0 E^2/2$ $u = 0.04425$ J/m³

(*c*) $U = uAd$ $U = 0.04425 \times 2 \times 10^{-3}$ J = 88.5 μJ

(*d*) $C = \epsilon_0 A/d$ $C = 8.85 \times 10^{-12} \times 2/10^{-3}$ F = 17.7 nF

(*e*) $U = \frac{1}{2}CV^2$ $U = 17.7 \times 10^{-9} \times 10^4/2$ J = 88.5 μJ = uAd

---

25* •   True or false:

(*a*) The equivalent capacitance of two capacitors in parallel equals the sum of the individual capacitances.

(*b*) The equivalent capacitance of two capacitors in series is less than the capacitance of either capacitor alone.

(*a*) True (*b*) True

29* •   Three capacitors are connected in a triangular network as shown in Figure 25-23. Find the equivalent capacitance across terminals *a* and *c*.

The series combination $C_1 + C_3$ has a capacitance $C_1 C_3/(C_1 + C_3)$. This is in parallel with C_2, so the total capacitance between *a* and *c* is $C = C_2 + C_1 C_3/(C_1 + C_3)$.

33* ••   For the circuit shown in Figure 25-24, find (*a*) the total equivalent capacitance between the terminals, (*b*) the charge stored on each capacitor, and (*c*) the total stored energy.

(*a*) $C_{equ} = C_1 C_2/(C_1 + C_2) + C_3$ $C_{equ} = (4 \times 15/19 + 12)$ μF = 15.2 μF

(*b*) $Q_3 = C_3 V$; $Q_1 = Q_2 = C_{series} V$ $Q_{12} = 2.4$ mC; $Q_4 = Q_{15} = (60/19) \times 200$ μC = 0.632 mC

(*c*) $U = \frac{1}{2}C_{equ}V^2$ $U = 0.304$ J

37* ••   In Figure 25-27, $C_1 = 2$ μF, $C_2 = 6$ μF, and $C_3 = 3.5$ μF. (*a*) Find the equivalent capacitance of this combination. (*b*) If the breakdown voltages of the individual capacitors are $V_1 = 100$ V, $V_2 = 50$ V, and $V_3 = 400$ V, what maximum voltage can be placed across points *a* and *b*?

(*a*) $C_{equ} = C_1 C_2/(C_1 + C_2) + C_3$ $C_{equ} = 5$ μF

(*b*) $V_1 = 3V_2$, and $V_3 = V_1 + V_2$; V_1 is critical $V_1 = 100$ V, then $V_3 = V_{max} = 133$ V

41* •   An electric field of 2×10^4 V/m exists between the plates of a circular parallel-plate capacitor that has a plate separation of 2 mm. (*a*) What is the voltage across the capacitor? (*b*) What plate radius is required if the stored charge is 10 μC?

(*a*) $V = Ed$ $V = 2 \times 10^4 \times 2 \times 10^{-3}$ V = 40 V

(*b*) $C = Q/V = \epsilon_0(\pi R^2)/d$; $R = (Qd/V\pi\epsilon_0)^{\frac{1}{2}}$ $R = 4.24$ m

---

45* •   A Geiger tube consists of a wire of radius 0.2 mm and length 12 cm and a coaxial cylindrical shell conductor of the same length and a radius of 1.5 cm. (*a*) Find the capacitance, assuming that the gas in the tube has a dielectric constant of 1. (*b*) Find the charge per unit length on the wire when the potential difference between the wire and shell is 1.2 kV.

(a) Use Equ. 25-11

$C = 1.55$ pF

(b) $Q/L = CV/L$

$Q/L = 15.5$ nC/m

49* •• A spherical capacitor has an inner sphere of radius R_1 with a charge of $+Q$ and an outer concentric spherical shell of radius R_2 with a charge of $-Q$. (a) Find the electric field and the energy density at any point in space. (b) Calculate the energy in the electrostatic field in a spherical shell of radius r, thickness dr, and volume $4\pi r^2\, dr$ between the conductors? (c) Integrate your expression from part (b) to find the total energy stored in the capacitor, and compare your result with that obtained using $U = \frac{1}{2}QV$.

(a) From Gauss's law, $E = kQ/r^2$ and $u = \frac{1}{2}\epsilon_0 E^2 = \frac{1}{2}k^2\epsilon_0 Q^2/r^4$.

(b) $dU = 4\pi r^2 u(r)dr = (kQ^2/2r^2)dr$.

(c) $U = \dfrac{kQ^2}{2}\displaystyle\int_{R_1}^{R_2}\dfrac{dr}{r^2} = \dfrac{kQ^2(R_2 - R_1)}{2R_1 R_2} = \dfrac{1}{2}Q^2\left(\dfrac{R_2 - R_1}{4\pi\epsilon_0 R_1 R_2}\right)$. Note that the quantity within the parentheses is $1/C$

(see Problem 48), so $U = \frac{1}{2}Q^2/C$.

53* •• A 100-pF capacitor and a 400-pF capacitor are both charged to 2.0 kV. They are then disconnected from the voltage source and are connected together, positive plate to positive plate and negative plate to negative plate. (a) Find the resulting potential difference across each capacitor. (b) Find the energy lost when the connections are made.

(a) Both C_1 and C_2 have the same potential difference $V = 2$ kV

(b) There is no energy loss $\Delta U = 0$

57* •• Work Problem 54 if the two capacitors are first connected in parallel across the 12-V battery and are then connected, with the positive plate of each capacitor connected to the negative plate of the other.

(a) Find Q_1 and Q_2; then $Q_f = Q_2 - Q_1$ $Q_1 = 48\ \mu C$; $Q_2 = 144\ \mu C$; $Q_f = 96\ \mu C$

For the parallel connection $(C_1 + C_2)V_f = Q$ $V_f = (96/16)$ V $= 6$ V

(b) $U = \frac{1}{2}CV^2$ $U_i = \frac{1}{2}(16\times144)\ \mu J = 1.15$ mJ; $U_f = 0.288$ mJ

61* • A parallel-plate capacitor is made by placing polyethylene ($\kappa = 2.3$) between two sheets of aluminum foil. The area of each sheet is 400 cm^2, and the thickness of the polyethylene is 0.3 mm. Find the capacitance.

$C = \kappa\epsilon_0 A/d$ $C = 2.71$ nF

65* •• A parallel-plate capacitor has plates separated by a distance s. The space between the plates is filled with two dielectrics, one of thickness $\frac{1}{4}s$ and dielectric constant κ_1, the other with thickness $\frac{3}{4}s$ and dielectric constant κ_2. Find the capacitance of this capacitor in terms of C_0, the capacitance with no dielectrics.

We can view the system as two capacitors in a series connection, C_1 of thickness $s/4$, C_2 of thickness $3s/4$. Then $C_1 = 4\kappa_1\epsilon_0 A/s$ and $C_2 = 4\kappa_2\epsilon_0 A/3s$. The series combination gives $C = C_1 C_2/(C_1 + C_2) = C_0[4\kappa_1\kappa_2/(3\kappa_1 + \kappa_2)]$, where $C_0 = \epsilon_0 A/s$.

69* •• Two parallel plates have charges Q and $-Q$. When the space between the plates is devoid of matter, the electric field is 2.5×10^5 V/m. When the space is filled with a certain dielectric, the field is reduced to 1.2×10^5 V/m. (a) What is the dielectric constant of the dielectric? (b) If $Q = 10$ nC, what is the area of the plates? (c) What is the total induced charge on either face of the dielectric?

(a) Use Equ. 25-22 $\kappa = 2.5/1.2 = 2.08$

(b) $A = Q/E_0\epsilon_0$ $A = 45.2$ cm^2

(c) Use Equ. 25-27; $Q_b/Q_f = \sigma_b/\sigma_f$ $Q_b = 5.20$ nC

73* •   True or false:

(a) The capacitance of a capacitor is defined as the total amount of charge it can hold.

(b) The capacitance of a parallel-plate capacitor depends on the voltage difference between the plates.

(c) The capacitance of a parallel-plate capacitor is proportional to the charge on its plates.

(a) False (b) False (c) False

77* •   The voltage across a parallel plate capacitor with plate separation 0.5 mm is 1200 V. The capacitor is disconnected from the voltage source and the separation between the plates is increased until the energy stored in the capacitor has been doubled. Determine the final separation between the plates.

Since Q and σ are constant, so is E. $U \propto AdE^2$, so to double U one must double d. $d_f = 1.0$ mm.

81* ••   Repeat Problem 80 if the region filled with dielectric of capacitor (a) is two-thirds of the volume between the plates.

For C_1 we have a series combination of $3\kappa C_0/2$ and $3C_0$, where C_0 is the capacitance of the parallel plate capacitor without the dielectric. Thus $C_1 = C_0[3\kappa/(\kappa + 2)]$. $C_2 = C_0(1 - x) + C_0\kappa x$, where x is the fraction containing the dielectric. Solving for x one obtains $x = 2/(\kappa + 2)$.

85* ••   Estimate the capacitance of a typical hot-air balloon.

Approximate the balloon by a sphere of radius $R = 3$ m. Then $C = R/k = 0.3$ nF.

89* ••   Suppose the capacitor of Problem 88 is connected to a constant voltage source of 20 V. How far should the dielectric slab be pulled so that the stored energy is reduced to half its initial value?

Now V is constant. $U_i = \frac{1}{2}(4C_0)V^2$, where C_0 is the capacitance without the dielectric. When the dielectric is pulled out we have two capacitors in parallel. Let $s = x/L$. Then $C_f = 4C_0(1 - s) + C_0 s$, and $U_f = \frac{1}{2}C_0V^2[4(1 - s) + s]$. Set $U_f = \frac{1}{2}U_i$ and solve for s: $s = 2/3$ and $x = 2L/3 = 6.67$ cm.

93* ••   A parallel-plate capacitor is filled with two dielectrics of equal size as shown in Figure 25-35. (a) Show that this system can be considered to be two capacitors of area $\frac{1}{2}A$ connected in parallel. (b) Show that the capacitance is increased by the factor $(\kappa_1 + \kappa_2)/2$.

(a) Since the capacitor plates are conductors the potentials are the same across the entire upper and lower plates. So the system is equivalent to two capacitors in parallel, each of area $A/2$.

(b) $C = \frac{1}{2}C_0\kappa_1 + \frac{1}{2}C_0\kappa_2 = \frac{1}{2}C_0(\kappa_1 + \kappa_2)$. $C/C_0 = (\kappa_1 + \kappa_2)/2$.

97* ••   A parallel-plate capacitor has a plate area of 1.0 m^2 and a plate separation distance of 0.5 cm. Completely filling the space between the conducting plates is a glass plate having a dielectric constant of 5.0. The capacitor is charged to a potential difference of 12.0 V and is then removed from its charging source. How much work is required to pull the glass plate out of the capacitor?

1. Find $Q = \kappa C_0 V$; $C_0 = \epsilon_0 A/d$ $Q = 5 \times 8.85 \times 10^{-12} \times 12/5 \times 10^{-3}$ C $= 0.106\ \mu$C

2. Find $\Delta U = (\frac{1}{2}Q^2/C_0)(1 - 1/\kappa)$ $\Delta U = W = 2.54\ \mu$J

101* •••   You are asked to construct a parallel-plate, air-gap capacitor that will store 100 kJ of energy. (a) What minimum volume is required between the plates of the capacitor? (b) Suppose you have developed a dielectric that can withstand 3×10^8 V/m and has a dielectric constant of 5. What volume of this dielectric between the plates of the capacitor is required for it to be able to store 100 kJ of energy?

(a) $U = \frac{1}{2}\epsilon_0 E^2 v$, where v denotes volume; $E = E_{max}$ $v = 2 \times 10^5/(9 \times 10^{12} \times 8.85 \times 10^{-12})$ m^3 = 2.51×10^3 m^3

(b) Use $E_{max} = 3 \times 10^8$; $u = \frac{1}{2}\kappa\epsilon_0 E_{max}^2$ $v = 5.02 \times 10^{-2}$ m^3

105* ••• A spherical weather balloon made of aluminized Mylar and filled with helium at atmospheric pressure can lift a payload of 0.2 kg. Determine the capacitance of the balloon. (Neglect the mass of the Mylar.)

1. Equate buoyant force to total weight $0.2g + 4\pi R^3 \rho_{He} g/3 = 4\pi R^3 \rho_{air} g/3$

2. Solve for R $R = [0.6/4\pi(1.293 - 0.1786)]^{1/3}$ m = 0.35 m

3. Find $C = R/k$ $C = 38.9$ pF

109* ••• Two identical, 10-μF parallel-plate capacitors are given equal charges of 100 μC each and are then removed from the charging source. The charged capacitors are connected by a wire between their positive plates and another wire between their negative plates. (a) What is the stored energy of the system? A dielectric having a dielectric constant of 3.2 is inserted between the plates of one of the capacitors such that it completely fills the region between the plates. (b) What is the final charge on each capacitor? (c) What is the final stored energy of the system?

(a) $U = 2(\frac{1}{2}Q_1^2/C_1)$ $U = 1$ mJ

(b) $C_{equ} = C_1(1 + \kappa)$; $Q = Q_i = 2Q_1$; find V, Q_1', Q_2' $Q = 200\ \mu$C; $V = 200/42$ V = 4.76 V;

$Q_1' = 47.6\ \mu$C, $Q_2' = 152.4\ \mu$C

(c) $U_f = \frac{1}{2}QV$ $U_f = 0.476$ mJ

113* ••• A variable air capacitor like the one shown in the photograph on page 759 has a capacitance that changes between 0.02 and 0.12 μF as the shaft is rotated through an angle of 180°. A voltage of 100 V is maintained between the capacitor plates. Initially the capacitor is in its minimum capacitance position. (a) How much work must be done to rotate the shaft to the maximum capacitance position? (b) The shapes of the plates are designed so that the capacitance is a linear function of rotation angle. How much torque must be applied to rotate the capacitor to hold it in the position corresponding to $C = 0.07\ \mu$F?

(a) $W = U_f - U_i = \frac{1}{2}V^2(C_f - C_i)$ $W = 0.5$ mJ

(b) $\tau = dW/d\theta = (dW/dC)(dC/d\theta)$; $C = C_0 + \alpha\theta$ $\tau = \frac{1}{2}V^2\alpha$, $\alpha = (0.1/\pi)\ \mu$F/rad; $\tau = 1.59 \times 10^{-4}$ N·m

CHAPTER 26

Electric Current and Direct-Current Circuits

1* • In our study of electrostatics, we concluded that there is no electric field within a conductor in electrostatic equilibrium. How is it that we can now discuss electric fields inside a conductor?

When a current flows, the charges are not in equilibrium. In that case, the electric field provides the force needed for the charge flow.

5* • In a certain electron beam, there are 5.0×10^6 electrons per cubic centimeter. Suppose the kinetic energy of each electron is 10.0 keV, and the beam is cylindrical, with a diameter of 1.00 mm. (*a*) What is the velocity of an electron in the beam? (*b*) Find the beam current.

(*a*) $v = \sqrt{2K/m_e}$; $K = 10^4 \times 1.6 \times 10^{-19}$ J $\qquad\qquad v = 5.93 \times 10^7$ m/s

(*b*) $I = nevA$; $A = \pi D^2/4$ $\qquad\qquad\qquad\qquad I = 37.2\ \mu A$

9* •• In a certain particle accelerator, a proton beam with a diameter of 2.0 mm constitutes a current of 1.0 mA. The kinetic energy of each proton is 20 MeV. The beam strikes a metal target and is absorbed by it. (*a*) What is the number n of protons per unit volume in the beam? (*b*) How many protons strike the target in 1.0 min? (*c*) If the target is initially uncharged, express the charge of the target as a function of time.

(*a*) 1. $I = neAv$; $v = \sqrt{2K/m_p}$ $\qquad\qquad v = \sqrt{\dfrac{40 \times 10^6 \times 1.6 \times 10^{-19}}{1.67 \times 10^{-27}}} = 6.19 \times 10^7$ m/s;

     2. $A = \pi D^2/4$; solve for n $\qquad\qquad A = \pi \times 10^{-6}$ m^2; $n = 3.21 \times 10^{13}$

(*b*) $N = nAvt$ $\qquad\qquad\qquad\qquad\qquad N = 3.75 \times 10^{17}$

(*c*) $Q = I \times t$ $\qquad\qquad\qquad\qquad\qquad\quad Q = (1.0\ \text{mC/s})t$

13* • Two wires of the same material with the same length have different diameters. Wire A has twice the diameter of wire B. If the resistance of wire B is R, then what is the resistance of wire A? (*a*) R (*b*) $2R$ (*c*) $R/2$ (*d*) $4R$ (*e*) $R/4$

(*e*)

17* •• Two cylindrical copper wires have the same mass. Wire A is twice as long as wire B. Their resistances are related by

(a) $R_A = 8R_B$.   (b) $R_A = 4R_B$.   (c) $R_A = 2R_B$.   (d) $R_A = R_B$.

(b)

21* •   A carbon rod with a radius of 0.1 mm is used to make a resistor. The resistivity of this material is 3.5×10^{-5} Ω·m. What length of the carbon rod will make a 10-Ω resistor?

$L = RA/\rho = \pi r^2 R/\rho$ $L = \pi \times 10^{-8} \times 10/3.5 \times 10^{-5}$ m = 8.98 mm

---

25* ••   A cylinder of glass 1 cm long has a resistivity of 10^{12} Ω·m. How long would a copper wire of the same cross-sectional area need to be to have the same resistance as the glass cylinder?

$L_{Cu} = L_{glass}(\rho_{glass}/\rho_{Cu})$ $L_{Cu} = 0.01(10^{12}/1.7 \times 10^{-8})$ m = 5.88×10^{17} m = 62.2 c·y

---

29* ••   A rubber tube 1 m long with an inside diameter of 4 mm is filled with a salt solution that has a resistivity of 10^{-3} Ω·m. Metal plugs form electrodes at the ends of the tube. (a) What is the resistance of the filled tube? (b) What is the resistance of the filled tube if it is uniformly stretched to a length of 2 m?

(a) $R = \rho L/A$ $R = 79.6$ Ω

(b) $L' = 2L, A' = A/2; R' = 4R$ $R' = 318$ Ω

33* •••   A semiconducting diode is a nonlinear device whose current I is related to the voltage V across the diode by $I = I_0(e^{eV/kT} - 1)$, where k is Boltzmann's constant, e is the magnitude of the charge on an electron, and T is the absolute temperature. If $I_0 = 10^{-9}$ A and $T = 293$ K, (a) what is the resistance of the diode for $V = 0.5$ V? (b) What is the resistance for $V = 0.6$ V?

(a), (b) $R = V/I = V/[I_0(e^{eV/kT} - 1)]$ For $V = 0.5$ V, $eV/kT = 19.785; R = 1.28$ Ω

                                For $V = 0.6$ V, $R = 0.0293$ Ω

37* •••   The space between two metallic coaxial cylinders of length L and radii a and b is completely filled with a material having a resistivity ρ. (a) What is the resistance between the two cylinders? (See the hint in Problem 36.) (b) Find the current between the two cylinders if $\rho = 30$ Ω·m, $a = 1.5$ cm, $b = 2.5$ cm, $L = 50$ cm, and a potential difference of 10 V is maintained between the two cylinders.

(a) Here the element of resistance is a cylindrical shell of thickness dr and cross sectional area $2\pi rL$. The elements of resistance are in series. Thus

$$R = \frac{\rho}{2\rho L}\int_a^b \frac{dr}{r} = \frac{\rho}{2\pi L}\ln(b/a)$$

(b) Evaluate R and $I = V/R$ $R = 4.88$ Ω; $I = 2.05$ A

41* ••   An electric space heater has a Nichrome heating element with a resistance of 8 Ω at 20°C. When 120 V are applied, the electric current heats the Nichrome wire to 1000°C. (a) What is the initial current drawn by the cold heating element? (b) What is the resistance of the heating element at 1000°C? (c) What is the operating wattage of this heater?

(a) $I = V/R$ $I = 120/8$ A = 15 A

(b) $R_{1000} = R_{20}[1 + \alpha(1000 - 20)]$ $R_{1000} = 8(1 + 0.392)$ Ω = 11.14 Ω

(c) At $t_C = 1000°C, P = V^2/R_{1000}$ $P = 1.29$ kW

45* •   A resistor carries a current I. The power dissipated in the resistor is P. What is the power dissipated if the same resistor carries current $3I$? (Assume no change in resistance.) $(a) P$ $(b) 3P$ $(c) P/3$ $(d) 9P$ $(e) P/9$

(d)

49* •   Find the power dissipated in a resistor connected across a constant potential difference of 120 V if its resistance is (a) 5 Ω and (b) 10 Ω.

$(a), (b)$ $P = V^2/R$ 　　　　　　　　　　　　　　　　　　$(a) P = 2.88$ kW; $(b) P = 1.44$ kW

53* •   A battery with 12-V emf has a terminal voltage of 11.4 V when it delivers a current of 20 A to the starter of a car. What is the internal resistance r of the battery?

$r = V_r/I$ 　　　　　　　　　　　　　　　　　　$r = 0.6/20\ \Omega = 0.03\ \Omega$

57* ••   Staying up late to study, and having no stove to heat water, you use a 200-W heater from the lab to make coffee throughout the night. If 90% of the energy produced by the heater goes toward heating the water in your cup, (a) how long does it take to heat 0.25 kg of water from 15 to 100°C? (b) If you fall asleep while the water is heating, how long will it take to boil away after it reaches 100°C?

(a) $0.9P\Delta t = mc\Delta T$; $\Delta t = mc\Delta T/0.9P$ 　　　　$\Delta t = (0.25{\times}85{\times}4.18{\times}10^3/180)$ s $= 491$ s ≈ 8.2 min

(b) $0.9P\Delta t = mL$; $\Delta t = mL/0.9P$ 　　　　　　$\Delta t = (0.25{\times}2257{\times}10^3/180)$ s $= 3135$ s ≈ 52.2 min

---

61* ••   A lightweight electric car is powered by ten 12-V batteries. At a speed of 80 km/h, the average frictional force is 1200 N. (a) What must be the power of the electric motor if the car is to travel at a speed of 80 km/h? (b) If each battery can deliver a total charge of 160 A·h before recharging, what is the total charge in coulombs that can be delivered by the 10 batteries before charging? (c) What is the total electrical energy delivered by the 10 batteries before recharging? (d) How far can the car travel at 80 km/h before the batteries must be recharged? (e) What is the cost per kilometer if the cost of recharging the batteries is 9 cents per kilowatt-hour?

(a) $P = Fv$ 　　　　　　　　　　　　　$P = (1200{\times}22.2)$ W $= 26.7$ kW

(b) $Q = It$ 　　　　　　　　　　　　　$Q = (160{\times}10{\times}3600)$ C $= 5.76$ MC

(c) $W = Q\mathcal{E}$ 　　　　　　　　　　　　$W = 69.1$ MJ

(d) $W = Fd$ 　　　　　　　　　　　　　$d = 57.6$ km

(e) Cost = $\$0.09(\mathcal{E}It)/1000$ 　　　　Cost/km = $(0.09{\times}120{\times}160/10^3)/57.6 = \$0.03$/km

---

65* ••   When two identical resistors are connected in series across the terminals of a battery, the power delivered by the battery is 20 W. If these resistors are connected in parallel across the terminals of the same battery, what is the power delivered by the battery? (a) 5 W (b) 10 W (c) 20 W (d) 40 W (e) 80 W

(e)

69* •   In Figure 26-48, the current in the 4-Ω resistor is 4 A. (a) What is the potential drop between a and b? (b) What is the current in the 3-Ω resistor?

(a) $V = IR$ 　　　　　　　　　　　　　$V = 16$ V

(b) $I_3 = V/(R_3 + R_5)$ 　　　　　　　　$I_3 = 2$ A

73* ••   Consider the equivalent resistance of two resistors R_1 and R_2 connected in parallel as a function of the ratio $x = R_2/R_1$. (a) Show that $R_{eq} = R_1x/(1 + x)$. (b) Sketch a plot of R_{eq} as a function of x.

(a) $R_2 = xR_1$; then for parallel combination

$R_{equ} = xR_1^2/(R_1 + xR_1) = xR_1/(1 + x)$

(b) $r = R_{equ}/R_1$ versus x is shown in the figure

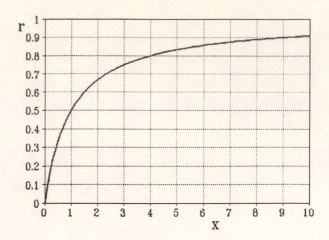

77* ••   A parallel combination of an 8-Ω resistor and an unknown resistor R is connected in series with a 16-Ω resistor and a battery. This circuit is then disassembled and the three resistors are then connected in series with each other and the same battery. In both arrangements, the current through the 8-Ω resistor is the same. What is the unknown resistance R?

1. For the first arrangement, find R_{equ} $R_{equ,1} = [16 + 8R/(R + 8)]\ \Omega = (24R + 128)/(R + 8)\ \Omega$

2. Write $I_{tot,1}$ and I_8 $I_{tot,1} = (R + 8)V/(24R + 128);\ I_{8,1} = I_{tot,1}R/(R + 8)$

3. For series arrangement, find R_{equ} and $I_{8,2}$ $I_{8,2} = V/(R + 24)$

4. Set $I_{8,1} = I_{8,2}$ and solve for R $R = \sqrt{128}\ \Omega = 11.3\ \Omega$

81* •   Kirchoff's loop rule follows from (a) conservation of charge. (b) conservation of energy. (c) Newton's laws. (d) Coulomb's law. (e) quantization of charge.

(b)

85* ••   In the circuit in Figure 26-57, the reading of the ammeter is the same with both switches open and both closed. Find the resistance R.

1. Find the current with both switches open $I = 1.5/450\ A = 3.33\ mA$

2. With both switches closed, 50 Ω is shorted; find $R_{equ} = [100R/(100 + R) + 300]\ \Omega;\ I_{tot} = 1.5/R_{equ};$

    $R_{equ}, I_{tot},$ and I_{100}; set $I_{100} = 3.33\ mA$ $I_{100} = 3.33 \times 10^{-3} = (1.5/R_{equ})[R/(100 + R)]$

3. Solve for R $R = 600\ \Omega$

89* ••   Two identical batteries, each with an emf $\mathcal{E}$ and an internal resistance r, can be connected across a resistance R either in series or in parallel. Is the power supplied to R greater when $R < r$ or when $R > r$?

If connected in series, $I = 2\mathcal{E}/(2r + R)$ and $P = I^2/R = 4\mathcal{E}^2R/(2r + R)^2$. If connected in parallel, the emf is $\mathcal{E}$ and the two internal resistances are in parallel and equivalent to $r/2$. Then $I = \mathcal{E}/(r/2 + R)$ and $P = \mathcal{E}^2R/(r/2 + R)^2 = 4\mathcal{E}^2R/(r + 2R)^2$. If $r = R$ both arrangements provide the same power to the load. In the parallel connection, the power is greater if $R < r$ and will be a maximum when $R = r/2$. For the series connection, the power to the load is greater if $R > r$ and is greatest when $R = 2r$.

93* •• You have two batteries, one with $\mathcal{E} = 9.0$ V and $r = 0.8$ Ω and the other with $\mathcal{E} = 3.0$ V and $r = 0.4$ Ω. (*a*) Show how you would connect the batteries to give the largest current through a resistor R. Find the current for (*b*) $R = 0.2$ Ω, (*c*) $R = 0.6$ Ω, (*d*) $R = 1.0$ Ω, and (*e*) $R = 1.5$ Ω.

Let $\mathcal{E}_1$ be the 9-V battery and r_1 its internal resistance of 0.8 Ω, $\mathcal{E}_2$ be the 3-V battery and r_2 its internal resistance of 0.4 Ω. If the two batteries are connected in series and then connected to the load resistance R, the current through R is $I_s = (\mathcal{E}_1 + \mathcal{E}_2)/(r_1 + r_2 + R) = 12/(1.2 + R)$ A.

Suppose the two batteries are connected in parallel and their terminals are then connected to R. Let I_1 be the current delivered by $\mathcal{E}_1$, I_2 be the current delivered by $\mathcal{E}_2$, and I_p the current through the load resistor R in the parallel connection.

1. Write the junction equations	$I_1 + I_2 = I_p$
2. Write two loop equations	$9 - 0.8I_1 - I_pR = 0;\ 3 - 0.4I_2 - I_pR = 0$
3. Solve for I_p	$I_p = 7.5/(0.4 + 1.5R)$ A
(*b*) Evaluate I_s and I_p for $R = 0.2$ Ω	$I_s = 8.57$ A; $I_p = 10.7$ A
(*c*) Evaluate I_s and I_p for $R = 0.6$ Ω	$I_s = 6.67$ A; $I_p = 5.77$ A
(*d*) Evaluate I_s and I_p for $R = 1.0$ Ω	$I_s = 5.45$ A; $I_p = 3.95$ A
(*e*) Evaluate I_s and I_p for $R = 1.5$ Ω	$I_s = 4.44$ A; $I_p = 2.83$ A

Note that for $R = 0.4$ Ω, $I_s = I_p = 7.5$ A. When $R < 0.4$ Ω, the parallel connection gives the larger current through R. When $R > 0.4$ Ω, the series connection gives the larger current through R.

97* ••• Find the current in each resistor of the circuit shown in Figure 26-66.

Let I_1 be the current in the 3-Ω resistor in series with the 8-V battery, I_2 the current delivered by the 12-V battery, I_3, directed down, be the current through the 2-Ω resistor, I_4, directed down, the current through the 4-Ω resistor, I_5, directed down, the current through the 3-Ω resistor, and I_6, directed to the right, the current through the 5-Ω resistor.

1. Write the junction equations	$I_3 = I_1 + I_2;\ I_2 = I_5 + I_6;\ I_3 = I_4 + I_5$
2. Write three loop equations	$8 - 3I_1 - 2I_3 - 4I_4 = 0;\ 12 - 2I_3 - 3I_5 = 0;$
	$12 + 3I_1 - 8 - 5I_6 = 0$
3. Solve for the currents	$I_1 = -0.0848$ A; $I_2 = 2.883$ A; $I_3 = 2.80$ A;
	$I_4 = 0.666$ A; $I_5 = 2.134$ A; $I_6 = 0.749$ A

101* •• A battery is connected to a series combination of a switch, a resistor, and an initially uncharged capacitor. The switch is closed at $t = 0$. Which of the following statements is true?

(*a*) As the charge on the capacitor increases, the current increases.

(*b*) As the charge on the capacitor increases, the voltage drop across the resistor increases.

(*c*) As the charge on the capacitor increases, the current remains constant.

(*d*) As the charge on the capacitor increases, the voltage drop across the capacitor decreases.

(*e*) As the charge on the capacitor increases, the voltage drop across the resistor decreases.

(*e*)

105* •• A 6-μF capacitor is charged to 100 V and is then connected across a 500-Ω resistor. (*a*) What is the initial charge on the capacitor? (*b*) What is the initial current just after the capacitor is connected to the resistor? (*c*) What is the time constant of this circuit? (*d*) How much charge is on the capacitor after 6 ms?

(a) $Q_0 = CV_0$ $Q_0 = 600\ \mu C$

(b) $I_0 = V_0/R$ $I_0 = 0.2$ A

(c) $\tau = RC$ $\tau = 3$ ms

(d) $Q(t) = Q_0 e^{-t/\tau}$ $Q = (600\ e^{-2})\ \mu C = 81.2\ \mu C$

109*•• A 1.6-μF capacitor, initially uncharged, is connected in series with a 10-kΩ resistor and a 5.0-V battery of negligible internal resistance. (a) What is the charge on the capacitor after a very long time? (b) How long does it take the capacitor to reach 99% of its final charge?

(a) $Q = CV$ $Q = 8.0\ \mu C$

(b) $\tau = RC$; $e^{-t/\tau} = 0.01$ $\tau = 0.016$ s; $t/\tau = \ln(100)$; $t = 73.7$ ms

---

113*•• In the steady state, the charge on the 5-μF capacitor in the circuit in Figure 26-69 is 1000 μC. (a) Find the battery current. (b) Find the resistances R_1, R_2, and R_3.

(a) 1. Find the current in the 10-Ω resistor $I_{10\Omega} = V_C/10$ A $= 200/10$ A $= 20$ A

   2. $I_{bat} = I_{10\Omega} + 5$ A $I_{bat} = 25$ A

(b) 1. Find $I_{5\Omega}$, I_{R3}, and I_{R1} $I_{5\Omega} = 10$ A, $I_{R3} = 15$ A, $I_{R1} = I_{bat} = 25$ A

   2. Write loop equations and solve for unknown $310 - 25R_1 - 5\times50 - 5\times10 = 0$; $R_1 = 0.4\ \Omega$

      resistors. $310 - 10 - 200 - 15R_3 = 0$; $R_3 = 6.67\Omega$

                                        $200 + 5R_2 = 5\times50$; $R_2 = 10\ \Omega$

117*••• For the circuit in Figure 26-71, (a) what is the initial battery current immediately after switch S is closed? (b) What is the battery current a long time after switch S is closed? (c) What is the current in the 600-Ω resistor as a function of time?

(a) $V_{C0} = 0$; $I_0 = \mathcal{E}/R_{200}$ $I_0 = 50/200$ A $= 0.25$ A

(b) $I_\infty = \mathcal{E}/R_{tot}$ $I_\infty = 50/800$ A $= 0.0625$ A

(c) 1. Let $R_1 = 200$-Ω resistor, $R_2 = 600$-Ω resistor, $I_1 = I_2 + I_3$ (1); $\mathcal{E} - R_1 I_1 - Q/C = 0$ (2);

    and I_1, I_2 be their currents, and I_3 = current into C $Q/C = R_2 I_2$ (3)

   2. Differentiate equation (2) with respect to time $R_1(dI_1/dt) = -(1/C)(dQ/dt) = -(1/C)I_3$

   3. Differentiate equation (3) with respect to time $(1/C)I_3 = R_2(dI_2/dt)$; $dI_1/dt = -(R_2/R_1)(dI_2/dt)$

   4. Use junction equation $dI_2/dt = (I_1 - I_2)/R_2 C$

   5. From equations (1) and (3) solve for I_1 $I_1 = (\mathcal{E} - R_2 I_2)/R_1$

   6. Write the differential equation for I_2 $dI_2/dt = (\mathcal{E}/R_1 R_2 C) - I_2(R_1 + R_2)/R_1 R_2 C$

   7. To solve, let $I_2(t) = a + be^{-t/\tau}$; substitute for I_2 $a = \mathcal{E}/(R_1 + R_2)$; $b = -a$; $\tau = R_1 R_2 C/(R_1 + R_2)$

     and solve for a, b, and τ using $I_2(0) = 0$ $I_2 = 0.0625(1 - e^{-t/\tau})$, where $\tau = 0.75$ ms

---

121*• A flash lamp is set off by the discharge of a capacitor that has been charged by a battery. Why not just connect the battery directly to the lamp?

The battery cannot deliver energy at the high rate required to light the flash.

125*•• In Figure 26-74, all three resistors are identical. The power dissipated is

(a) the same in R_1 as in the parallel combination of R_2 and R_3.

(b) the same in R_1 and R_2.

(c) greatest in R_1.

(d) smallest in R_1.

(c)

129*• A 10.0-Ω resistor is rated as being capable of dissipating 5.0 W of power. (a) What maximum current can this resistor tolerate? (b) What voltage across this resistor will produce the maximum current?

(a) $I^2R = P$ $I_{max} = 0.707$ A

(b) $V = IR$ $V = 7.07$ V

133*•• A 16-gauge copper wire insulated with rubber can safely carry a maximum current of 6 A. (a) How great a potential difference can be applied across 40 m of this wire? (b) Find the electric field in the wire when it carries a current of 6 A. (c) Find the power dissipated in the wire when it carries a current of 6 A.

(a) Find $R = \rho L/A$; see Table 26-2 for A $R = 1.7 \times 10^{-8} \times 40/1.31 \times 10^{-6}$ $\Omega = 0.519$ Ω

       $V = IR$ $V = 3.11$ V

(b) $E = V/L$ $E = 0.0779$ V/m

(c) $P = I^2R$ $P = 18.7$ W

137*•• The capacitors in the circuit in Figure 26-75 are initially uncharged. (a) What is the initial value of the battery current when switch S is closed? (b) What is the battery current after a long time? (c) What are the final charges on the capacitors?

(a) Since $V_C = 0$ for both capacitors, the resistors are $R_{equ} = (7.5 \times 12/19.5 + 10)$ $\Omega = 14.6$ Ω

     effectively in parallel. Find R_{equ} of circuit and I_0. $I_0 = 3.42$ A

(b) Now $I_C = 0$, and resistors are in series. Find I_∞. $I_\infty = 50/52$ A $= 0.962$ A

(c) Find V_C for 10 μF and 5 μF capacitors   For both capacitors, $V_C = 27 \times 0.962$ V $= 26.0$ V

     $Q = CV$ For $C = 10$ μF, $Q = 260$ μC; for $C = 5\mu$F, $Q = 130$ μC

141*•• You are given n identical cells, each with emf $\mathcal{E}$ and internal resistance $r = 0.2$ Ω. When these cells are connected in parallel to form a battery, and a resistance R is connected to the battery terminal, the current through R is the same as when the cells are connected in series and R is attached to the terminals of that battery. Find the value of the resistor R.

1. Write I for series connection $I_s = n\mathcal{E}/(nr + R)$

2. Write I for parallel connection $I_p = \mathcal{E}/(r/n + R)$

3. Set $I_s = I_p$ and solve for R with $r = 0.2$ Ω $R = 0.2$ Ω

145*•• The belt of a Van de Graaff generator carries a surface charge density of 5 mC/m². The belt is 0.5 m wide and moves at 20 m/s. (a) What current does it carry? (b) If this charge is raised to a potential of 100 kV, what is the minimum power of the motor needed to drive the belt?

(a) $I = dQ/dt = \sigma w\, dx/dt = \sigma wv$ $I = (5 \times 10^{-3} \times 0.5 \times 20)$ A $= 50$ mA

(b) $P = IV$ $P = 5$ kW

149*••• If the capacitor in the circuit in Figure 26-70 is replaced by a 30-Ω resistor, what currents flow through the resistors?

1. Write the junction equations; currents are down, and in 30-Ω resistor to the right

$I_{10} + I_{40} = I_{80} + I_{20}$; $I_{10} = I_{30} + I_{80}$

2. Write the loop equations

$36 - 10I_{10} - 80I_{80} = 0$; $36 - 40I_{40} - 20I_{20} = 0$;

$10I_{10} + 30I_{30} - 40I_{40} = 0$

3. Solve the set of five linear equations

$I_{10} = 0.740$ A; $I_{40} = 0.472$ A; $I_{30} = 0.383$ A;

$I_{80} = 0.357$ A; $I_{20} = 0.855$ A

153*••• In the circuit in Figure 26-80, the capacitors are initially uncharged. Switch S_2 is closed and then switch S_1 is closed. (*a*) What is the battery current immediately after S_1 is closed? (*b*) What is the battery current a long time after both switches are closed? (*c*) What is the final voltage across C_1? (*d*) What is the final voltage across C_2? (*e*) Switch S_2 is opened again after a long time. Give the current in the 150-Ω resistor as a function of time.

(*a*) At $t = 0$ the capacitor voltages are zero. So at $t = 0$, effectively, $R = 100\ \Omega$. Find $I_{bat}(0)$.

$I_{bat} = 12/100$ A = 0.12 A

(*b*) For $t = \infty$, $R_{equ} = 300\ \Omega$

$I_{bat} = 12/300$ A = 0.040 A

(*c*) $V_{C1} = \mathcal{E} - 100I_{bat}$

$V_{C1} = (12 - 4.0)$ V = 8.0 V

(*d*) $V_{C2} = 150I_{bat}$

$V_{C2} = 6.0$ V

(*e*) $I = I_0 e^{-t/\tau}$; $\tau = RC$

$I = [0.040 \exp(-t/7.5 \times 10^{-3})]$ A

CHAPTER 27

The Microscopic Theory of Electrical Conduction

1* •   In the classical model of conduction, the electron loses energy on average in a collision because it loses the drift velocity it had picked up since the last collision. Where does this energy appear?

The energy lost by the electrons in collision with the ions of the crystal lattice appears as Joule heat (I^2R).

5* •   The density of aluminum is 2.7 g/cm³. How many free electrons are present per aluminum atom?

n_e = electrons/atom = $nM/\rho N_A$ $n_e = \dfrac{18.1 \times 10^{22} \times 26.98}{2.7 \times 6.02 \times 10^{23}} = 3.00$ electrons/atom

---

9* •   Calculate the Fermi energy for (a) Al, (b) K, and (c) Sn using the number densities given in Table 27-1.

(a), (b), (c) Use Equ. 27-15b   (a) $E_F = 0.365(181)^{2/3}$ eV = 11.7 eV;

(b) $E_F = 2.12$ eV; (c) $E_F = 10.2$ eV

13* ••   The bulk modulus B of a material can be defined by

$$B = -V\frac{\partial P}{\partial V}$$

(a) Use the ideal-gas relation $PV = \frac{2}{3}NE_{av}$ and Equations 27-15 and 27-16 to show that

$$P = \frac{2NE_F}{5V} = CV^{-5/3}$$

where C is a constant independent of V. (b) Show that the bulk modulus of the Fermi electron gas is therefore

$$B = \frac{5}{3}P = \frac{2NE_F}{3V}$$

(c) Compute the bulk modulus in newtons per square meter for the Fermi electron gas in copper and compare your result with the measured value of 140×10^9 N/m².

(a) From Problem 12, $P = (2/3)(N/V)E_{av} = (2/5)(N/V)E_F$. But E_F is proportional to $V^{-2/3}$ so $P = CV^{-5/3}$, where C is a constant.

(b) $B = -(1/V)(dP/dV) = (5/3)CV^{-5/3} = (5/3)P = (2/3)(N/V)E_F$.

(c) $B = 63.6 \times 10^9$ N/m² ≈ $0.5B_{exp}$

17* •   When the temperature of pure copper is lowered from 300 K to 4 K, its resistivity drops by a much greater factor than that of brass when it is cooled in the same way. Why?

The resistivity of brass at 4 K is almost entire due to the "residual resistance," the resistance due to impurities and other imperfections of the crystal lattice. In brass, the zinc ions act as impurities in copper. In pure copper, the resistivity at 4 K is due to its residual resistance which is very low if the copper is very pure.

21* •   Insulators are poor conductors of electricity because

(a) there is a small energy gap between the valence band and the next higher band where electrons can exist.

(b) there is a large energy gap between the full valence band and the next higher band where electrons can exist.

(c) the valence band has a few vacancies for electrons.

(d) the valence band is only partly full.

(e) None of these is correct.

(b)

25* ••   A photon of wavelength 3.35 μm has just enough energy to raise an electron from the valence band to the conduction band in a lead sulfide crystal. (a) Find the energy gap between these bands in lead sulfide. (b) Find the temperature T for which kT equals this energy gap.

(a) $E_g = hc/\lambda$ $\qquad\qquad\qquad\qquad\qquad$ $E_g = 1240/3350$ eV $= 0.370$ eV

(b) $T = E_g/k$; $k = 8.62\times10^{-5}$ eV/K $\qquad$ $T = 4.29\times10^3$ K

29* ••   Show that at $E = E_F$, the Fermi factor is $F = 0.5$.

For $E = E_F$, $e^{(E - E_F)/kT} = e^0 = 1$. Consequently, $f(E_F) = \tfrac{1}{2}$

33* ••   Carry out the integration $E_{av} = (1/N)\int_0^{E_F} Eg(E)\ dE$ to show that the average energy at $T = 0$ is $\tfrac{3}{5}E_F$.

$$\frac{1}{N}\int_0^{E_F} Eg(E)\,dE = \frac{3}{2}E_F^{-3/2}\int_0^{E_F} E^{3/2}\,dE = \frac{3}{2}E_F^{-3/2}\frac{2}{5}E_F^{5/2} = \frac{3}{5}E$$

37* ••   In an intrinsic semiconductor, the Fermi energy is about midway between the top of the valence band and the bottom of the conduction band. In germanium, the forbidden energy band has a width of 0.7 eV. Show that at room temperature the distribution function of electrons in the conduction band is given by the Maxwell-Boltzmann distribution function.

$n(E) = g(E)f(E)$. In this case, $\exp[(E - E_F)/kT] = \exp[(E - E_g/2)/kT] \gg 1$ so the $f(E) = e^{E_g/2kT}e^{-E/kT}$. So, using Equ. 27-30, we have

$$n(E) = \frac{3N}{2}E_F^{-3/2}e^{E_g/2kT}E^{1/2}e^{-E/kT}$$

There is generally an additional temperature dependence that arises from the fact that E_F depends on T. At room temperature, $\exp[(E - E_g/2)/kT] \geq \exp(0.35/0.0259) = 7.4\times10^5$, so the approximation leading to the Boltzmann distribution is justified.

41* •••   When a star with a mass of about twice that of the sun exhausts its nuclear fuel, it collapses to a neutron star, a dense sphere of neutrons of about 10 km diameter. Neutrons are spin-$\tfrac{1}{2}$ particles and, like electrons, are subject to the exclusion principle. (*a*) Determine the neutron density of such a neutron star. (*b*) Find the Fermi energy of the neutron distribution.

(*a*) Find N/V; $N = M/m_n$; $V = \pi D^3/6$ $N/V = 4.57 \times 10^{45}$ m^{-3}

(*b*) Use Equ. 27-15*a*, replacing m_e by m_n $E_F = 8.77 \times 10^{-11}$ J = 548 MeV .

45* •   Calculate the number density of free electrons for (*a*) Mg ($\rho = 1.74$ g/cm^3) and (*b*) Zn ($\rho = 7.1$ g/cm^3), assuming two free electrons per atom, and compare your results with the values listed in Table 27-1.

(a), (b) $n = 2N_A\rho/M$ (*a*) $n = 2 \times 6.02 \times 10^{23} \times 1.74/24.3 = 8.62 \times 10^{22}$

The calculated values agree well with Table 27-1   (*b*) $n = 13.1 \times 10^{22}$

CHAPTER 28

The Magnetic Field

1* • When a cathode-ray tube is placed horizontally in a magnetic field that is directed vertically upward, the electrons emitted from the cathode follow one of the dashed paths to the face of the tube in Figure 28-30. The correct path is ____. (*a*) 1 (*b*) 2 (*c*) 3 (*d*) 4 (*e*) 5

(*b*)

5* • A uniform magnetic field of magnitude 1.48 T is in the positive *z* direction. Find the force exerted by the field on a proton if the proton's velocity is (*a*) $v = 2.7$ Mm/s i, (*b*) $v = 3.7$ Mm/s j, (*c*) $v = 6.8$ Mm/s k, and (*d*) $v = 4.0$ Mm/s $i + 3.0$ Mm/s j.

(*a*), (*b*), (*c*), (*d*) Use Equ. 28-1

(*a*) $F = 0.639$ pN $i \times k = -0.639$ pN j (*b*) $F = 0.876$ pN i

(*c*) $F = 0$ (*d*) $F = 0.71$ pN $i - 0.947$ pN j

9* • What is the force (magnitude and direction) on an electron with velocity $v = (2i - 3j) \times 10^6$ m/s in a magnetic field $B = (0.8i + 0.6j - 0.4k)$ T?

Use Equ. 28-4

$F = -0.192$ pN $i - 0.128$ pN $j - 0.576$ pN k; $F = 0.621$ pN

13* •• A current-carrying wire is bent into a semicircular loop of radius *R* that lies in the *xy* plane. There is a uniform magnetic field $B = Bk$ perpendicular to the plane of the loop (Figure 28-32). Show that the force acting on the loop is $F = 2IRBj$.

With the current in the direction indicated and the magnetic field in the *z* direction, pointing out of the plane of the paper, the force is in the radial direction. On an element of length $d\ell$ the force is $dF = BIR \, d\theta$ with *x* and *y* components $dF_x = BIR \cos \theta \, d\theta$ and $dF_y = BIR \sin \theta \, d\theta$. By symmetry, the *x* component of the force is zero.

$$F_y = \int_0^\pi BIR \sin \theta \, d\theta = 2IBR.$$

17* • True or false: The magnetic force does not accelerate a particle because the force is perpendicular to the velocity of the particle.

False

21* •   An electron from the sun with a speed of 1×10^7 m/s enters the earth's magnetic field high above the equator where the magnetic field is 4×10^{-7} T. The electron moves nearly in a circle except for a small drift along the direction of the earth's magnetic field that will take it toward the north pole. (*a*) What is the radius of the circular motion? (*b*) What is the radius of the circular motion near the north pole where the magnetic field is 2×10^{-5} T?

(*a*), (*b*) Use Equ. 28-6   (*a*) $r = 142$ m   (*b*) $r = 2.85$ m

---

25* ••   A beam of particles with velocity *v* enters a region of uniform magnetic field *B* that makes a small angle θ with *v*. Show that after a particle moves a distance $2\pi(m/qB)v \cos\theta$ measured along the direction of *B*, the velocity of the particle is in the same direction as it was when it entered the field.

The particle's velocity has a component v_1 parallel to *B* and a component v_2 normal to *B*. $v_1 = v \cos\theta$ and is constant. $v_2 = v \sin\theta$; the magnetic force due to this velocity component is qBv_2 and results in a circular motion perpendicular to *B*. The period of that circular motion is given by Equ. 28-7. At the end of one period, v_2 is the same as at the start of the period. In that time, the particle has moved a distance $v_1 T = (v \cos\theta)(2\pi m/qB)$ in the direction of *B*.

29* •   A velocity selector has a magnetic field of magnitude 0.28 T perpendicular to an electric field of magnitude 0.46 MV/m. (*a*) What must the speed of a particle be for it to pass through undeflected? What energy must (*b*) protons and (*c*) electrons have to pass through undeflected?

(*a*) Use Equ. 28-9 $v = (0.46 \times 10^6/0.28)$ m/s $= 1.64 \times 10^6$ m/s

(*b*) $K = \frac{1}{2}m_p v^2$ $K = 2.25 \times 10^{-15}$ J $= 14.1$ keV

(*c*) $K = \frac{1}{2}m_e v^2$ $K = 7.68$ eV

33* ••   A singly ionized ^{24}Mg ion (mass 3.983×10^{-26} kg) is accelerated through a 2.5-kV potential difference and deflected in a magnetic field of 557 G in a mass spectrometer. (*a*) Find the radius of curvature of the orbit for the ion. (*b*) What is the difference in radius for ^{26}Mg and ^{24}Mg ions? (Assume that their mass ratio is 26/24.)

(*a*) $r = \sqrt{\dfrac{2m\Delta V}{qB^2}}$; evaluate for ^{24}Mg $r_{24} = 63.5$ cm

(*b*) $r_{26} = r_{24}\sqrt{26/24}$; evaluate $\Delta r = r_{26} - r_{24}$ $\Delta r = 63.5(\sqrt{26/24} - 1)$ cm $= 2.59$ cm

---

37* ••   A cyclotron for accelerating protons has a magnetic field of 1.4 T and a radius of 0.7 m. (*a*) What is the cyclotron frequency? (*b*) Find the maximum energy of the protons when they emerge. (*c*) How will your answers change if deuterons, which have the same charge but twice the mass, are used instead of protons?

(*a*) Use Equ. 28-8 $f = 21.3$ MHz

(*b*) Use Equ. 28-13 $K = 7.36 \times 10^{-12}$ J $= 46$ MeV

(*c*) From Equs. 28-8 and 28-13, K and $f \propto 1/m$ $f_d = 10.7$ MHz; $K_d = 23$ MeV

41* •   What orientation of a current loop gives maximum torque?

The normal to the plane of the loop should be perpendicular to *B*.

45* •   Repeat Problem 44 if the wire is bent into an equilateral triangle of sides 8 cm.

(*a*) $\tau = \mu \times B$; $\mu = \pm IA\,k$; $B = B\,k$ $\tau = 0$

(*b*) $B = B\,i$ $\tau = \pm 2.08 \times 10^{-3}$ N·m j

49* •   The SI unit for the magnetic moment of a current loop is A·m². Use this to show that 1 T = 1 N/A·m.

Since $[\tau] = [\mu][B]$, $[B] = [\tau]/[\mu] = $ N·m/A·m² = N/A·m

53* ••   A particle of charge q and mass m moves in a circle of radius r and with angular velocity ω. (a) Show that the average current is $I = q\omega/2\pi$ and that the magnetic moment has the magnitude $\mu = \frac{1}{2}q\omega r^2$. (b) Show that the angular momentum of this particle has the magnitude $L = mr^2\omega$ and that the magnetic moment and angular momentum vectors are related by $\mu = (q/2m)L$.

(a) $I = \Delta q/\Delta t = q/T = qf = q\omega/2\pi$. $\mu = IA = (q\omega/2\pi)(\pi r^2) = q\omega r^2/2$

(b) The moment of inertia of the particle is mr^2, and so $L = mr^2\omega$. Both μ and L point in the direction ω, so $\mu = (q/2m)L$.

57* •••   A nonconducting rod of mass M and length ℓ has a uniform charge per unit length λ and rotates with angular velocity ω about an axis through one end and perpendicular to the rod. (a) Consider a small segment of the rod of length dx and charge $dq = \lambda\,dx$ at a distance x from the pivot (Figure 28-37). Show that the magnetic moment of this segment is $\frac{1}{2}\lambda\omega x^2\,dx$. (b) Integrate your result to show that the total magnetic moment of the rod is $\mu = \frac{1}{6}\lambda\omega\ell^3$. (c) Show that the magnetic moment μ and angular momentum L are related by $\mu = (Q/2M)L$, where Q is the total charge on the rod.

(a) The area enclosed by the rotating element of charge is πx^2. The time required for one revolution is $1/f = 2\pi/\omega$. The average current element is then $dI = \lambda\,dx\,\omega/2\pi$ and $d\mu = A\,dI = \frac{1}{2}\lambda\omega x^2 dx$.

(b) $\mu = \dfrac{1}{2}\lambda\omega\displaystyle\int_0^\ell x^2 dx = \dfrac{1}{6}\lambda\omega\ell^3$

(c) The angular momentum $L = I\omega$, where I is the moment of inertia of the rod, $I = (1/3)M\ell^2$. The total charge carried by the rod is $Q = \lambda\ell$. Thus $\mu = (Q/2M)L$. Moreover, since ω and $L = I\omega$ point in the same direction, $\mu = (Q/2M)L$.

61* •••   A solid cylinder of radius R and length L carries a uniform charge density $+\rho$ between $r = 0$ and $r = R_s$ and an equal charge density of opposite sign, $-\rho$, between $r = R_s$ and $r = R$. What must be the radius R_s so that on rotation of the cylinder about its axis the magnetic moment is zero?

For the solid cylinder of radius R_s, $Q_+ = \pi\rho R_s^2 L$ and $L = I\omega = \frac{1}{2}MR_s^2\omega$. Hence $\mu_+ = (Q_+/2M)L = \pi\rho L R_s^4\omega/4$.

For the cylindrical shell, $Q_- = -\pi\rho L(R^2 - R_s^2)$ and $L = I\omega = \frac{1}{2}M(R_s^2 + R^2)\omega$. Hence $\mu_- = -\pi\rho L(R^4 - R_s^4)\omega/4$.

Setting $\mu_+ + \mu_- = 0$ and solving for R_s one obtains $R_s = R/2^{1/4} = 0.841R$.

65* •••   A solid sphere of radius R carries a uniform charge density, $+\rho_0$, between $r = 0$ and $r = \frac{1}{2}R$ and an equal charge density of opposite sign, $-\rho_0$, between $r = \frac{1}{2}R$ and $r = R$. The sphere rotates about its diameter with angular velocity ω. Derive an expression for the magnetic moment of this rotating sphere.

For the inner sphere of radius R_i, $Q_i = 4\pi\rho_0 R_i^3/3$ and $L = 2MR_i^2\omega/5$. Hence $\mu_i = 4\pi\rho_0 R_i^5\omega/15$. For the outer spherical shell, the charge is $Q_o = -(4\pi\rho_0/3)(R^3 - R_i^3)$. If ρ is the mass density, then $M_o = (4\pi\rho/3)(R^3 - R_i^3)$ and the moment of inertia of the spherical shell is $I = (2/5)(4\pi\rho/3)(R^5 - R_i^5)$. Using the general result $\mu = (Q/2M)L$ one obtains $\mu_o = -(4\pi\rho_0/15)(R^5 - R_i^5)\omega$. We now set $R_i = R/2$ and $\mu = \mu_i + \mu_o$ and obtain $\mu = -\pi\rho_0 R^5\omega/4$.

69* •• Because blood contains charged ions, moving blood develops a Hall voltage across the diameter of an artery. A large artery with a diameter of 0.85 cm has a flow speed of 0.6 m/s. If a section of this artery is in a magnetic field of 0.2 T, what is the potential difference across the diameter of the artery?

$V_H = v_d B w$ $V_H = 1.02$ mV

73* • True or false:

(a) The magnetic force on a moving charged particle is always perpendicular to the velocity of the particle.

(b) The torque on a magnet tends to align the magnetic moment in the direction of the magnetic field.

(c) A current loop in a uniform magnetic field behaves like a small magnet.

(d) The period of a particle moving in a circle in a magnetic field is proportional to the radius of the circle.

(e) The drift velocity of electrons in a wire can be determined from the Hall effect.

(a) True (b) True (c) True (d) False (e) True

77* • A positively charged particle is moving northward in a magnetic field. The magnetic force on the particle is toward the northeast. What is the direction of the magnetic field? (a) Up (b) West (c) South (d) Down (e) This situation cannot exist.

(e)

81* • A long wire parallel to the x axis carries a current of 6.5 A in the positive x direction. There is a uniform magnetic field $B = 1.35$ T j. Find the force per unit length on the wire.

Use Equ. 28-4 $F = 8.775$ N k

85* •• A particle of mass m and charge q enters a region where there is a uniform magnetic field B along the x axis. The initial velocity of the particle is $v = v_{0x} i + v_{0y} j$ so the particle moves in a helix. (a) Show that the radius of the helix is $r = mv_{0y}/qB$. (b) Show that the particle takes a time $t = 2\pi m/qB$ to make one orbit around the helix.

(a) Since $B = B i$, $v_0 \times B = v_{0y} B k$; i.e., $v_x = v_{0x}$ and motion on the direction of the magnetic field is not affected by the field. In the plane perpendicular to i the motion is as described in Section 28-2, and the radius of the circular path is given by Equ. 28-6 with $v = v_{0y}$, i.e., $r = mv_{0y}/qB$.

(b) The time for one complete orbit is given by Equ. 28-7, i.e., $t = 2\pi m/qB$.

89* •• A conducting wire is parallel to the y axis. It moves in the positive x direction with a speed of 20 m/s in a magnetic field $B = 0.5$ T k. (a) What are the magnitude and direction of the magnetic force on an electron in the conductor? (b) Because of this magnetic force, electrons move to one end of the wire leaving the other end positively charged, until the electric field due to this charge separation exerts a force on the electrons that balances the magnetic force. Find the magnitude and direction of this electric field in the steady state. (c) Suppose the moving wire is 2 m long. What is the potential difference between its two ends due to this electric field?

(a) Use Equ. 28-1; $q = -1.6\times10^{-19}$ C $F = 1.6\times10^{-18}$ N j

(b) At steady state, $qE + F = 0$ $E = 10$ V/m j

(c) $\Delta V = E \Delta x$ $\Delta V = 20$ V

93* ••• A circular loop of wire with mass M carries a current I in a uniform magnetic field. It is initially in equilibrium with its magnetic moment vector aligned with the magnetic field. The loop is given a small twist about a diameter and then released. What is the period of the motion? (Assume that the only torque exerted on the loop is due to the magnetic field.)

$\tau = -\mu B \sin \theta \approx -IAB\theta = -\pi R^2 IB\theta = MR^2 (d^2\theta/dt^2)$. This is the differential equation for a SHO, and comparison with Equs. 14-2 and 14-12 shows that $T = 2\pi\sqrt{M/(\pi I B)}$.

CHAPTER **29**

Sources of the Magnetic Field

1* • Compare the directions of the electric and magnetic forces between two positive charges, which move along parallel paths (*a*) in the same direction, and (*b*) in opposite directions.

(*a*) The electric forces are repulsive; the magnetic forces are attractive (the two charges moving in the same direction act like two currents in the same direction).

(*b*) The electric forces are again repulsive; the magnetic forces are also repulsive.

5* • An electron orbits a proton at a radius of 5.29×10^{-11} m. What is the magnetic field at the proton due to the orbital motion of the electron?

1. Determine the speed of the electron; $mv^2/r = ke^2/r^2$ $v = e\sqrt{k/mr}$

2. Use Equ. 29-1; $B = (\mu_0 e^2/4\pi r^2)\sqrt{k/mr}$; evaluate B $B = 12.5$ T

9* • For the current element in Problem 8, find the magnitude and direction of dB at $x = 0$, $y = 3$ m, $z = 4$ m.

Use Equ. 29-3; here $r = 3$ m $j + 4$ m k, $r = 5$ m $dB = -9.6 \times 10^{-12}$ T i

---

13* • A single-turn, circular loop of radius 10.0 cm is to produce a field at its center that will just cancel the earth's magnetic field at the equator, which is 0.7 G directed north. Find the current in the loop and make a sketch showing the orientation of the loop and the current.

From Equ. 29-5, with $x = 0$, $B = \mu_0 I/2R$; find I $I = 2BR/\mu_0 = 2 \times 7 \times 10^{-5} \times 0.1/4\pi \times 10^{-7}$ A $= 11.1$ A

The loop is shown in the figure. It is in the vertical plane and normal to the north direction. The current is to the west at the top of the loop, to the east at the bottom of the loop.

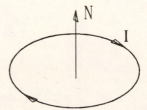

17* •• Two coils that are separated by a distance equal to their radius and that carry equal currents such that their axial fields add are called Helmholtz coils. A feature of Helmholtz coils is that the resultant magnetic field between the coils is very uniform. Let $R = 10$ cm, $I = 20$ A, and $N = 300$ turns for each coil. Place one coil in the yz plane with its center at the origin and the other in a parallel plane at $x = 10$ cm. (a) Calculate the resultant field B_x at $x = 5$ cm, $x = 7$ cm, $x = 9$ cm, and $x = 11$ cm. (b) Use your results and the fact that B_x is symmetric about the midpoint of the coils to sketch B_x versus x. (See also Problem 18.)

(a) For the coil centered at $x = 0$, $B_x(x)$ is given by Equ. 29-5 multiplied by N, the number of turns. For the coil centered at $x = R$, $B_x(x) = (\mu_0 R^2 NI/2)[(x - R)^2 + R^2]^{-3/2}$.

The total field is therefore

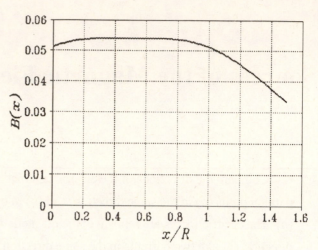

$$B_x(x) = \frac{\mu_0 NIR^2}{2}\left(\frac{1}{(x^2 + R^2)^{3/2}} + \frac{1}{[(x - R)^2 + R^2]^{3/2}}\right)$$

Evaluate $B_x(x)$ for $N = 300$, $R = 0.1$ m, and $I = 20$ A at $x = 0.05$ m, 0.07 m, 0.09 m, and 0.11 m. One obtains:
$B_x(0.05) = 0.0540$ T, $B_x(0.07) = 0.0539$ T,
$B_x(0.09) = 0.0526$ T, $B_x(0.11) = 0.0486$ T.

(b) $B(x)$ versus x/R is shown in the figure. Note that B is nearly constant between $x/R = 0.3$ and 0.7.

21* • A wire carries an electrical current straight up. What is the direction of the magnetic field due to the wire a distance of 2 m north of the wire? (a) North (b) East (c) West (d) South (e) Upward
(c)

25* • Sketch B_z versus y for points on the y axis when both currents are in the negative x direction.

$$B_z(y) = \frac{\mu_0 I}{2\pi}\left(\frac{1}{0.06 - y} + \frac{1}{-0.06 - y}\right)$$

The figure shows B_y as a function of y. Here B is in T, and y in m.

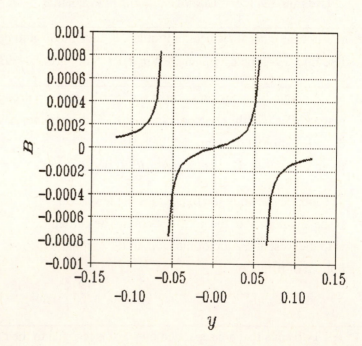

29* •   Find the magnitude of the force per unit length exerted by one wire on the other.

Use Equs. 29-12 and 28-5; $F/\ell = \mu_0 I^2/2\pi R$ $F/\ell = 6.67\times10^{-4}$ N/m

---

33* ••   Three long, parallel, straight wires pass through the corners of an equilateral triangle of sides 10 cm as shown in Figure 29-41, where a dot means that the current is out of the paper and a cross means that it is into the paper. If each current is 15.0 A, find (a) the force per unit length on the upper wire, and (b) the magnetic field B at the upper wire due to the two lower wires.

(a) The forces on the upper wire are away from and $F_i/\ell = \mu_0 I^2/2\pi R = 4.5\times10^{-4}$ N/m;

directed along the line to the lower wires $F_y/\ell = 2(F_i/\ell)\sin 60° = 7.79\times10^{-4}$ N/m

(b) By symmetry, $B_y = 0$; use $F_y = B_x I\ell$ $B = B_x i$; $B_x = F_y/I\ell = 5.20\times10^{-5}$ T

37* ••   Three very long, parallel wires are at the corners of a square, as shown in Figure 29-42. They each carry a current of magnitude I. Find the magnetic field B at the unoccupied corner of the square when (a) all the currents are into the paper, (b) I_1 and I_3 are in and I_2 is out, and (c) I_1 and I_2 are in and I_3 is out.

(a) Use Equ. 29-12; let $C = \mu_0 I/2\pi L$ $B_1 = -Cj$, $B_2 = (C/\sqrt{2})(i - j)/\sqrt{2}$, $B_3 = Ci$

$B = B_1 + B_2 + B_3$ $B = (3C/2)(i - j) = (3\mu_0 I/4\pi L)(i - j)$

(b) Proceed as in part (a); note that B_2 is reversed $B_1 = -Cj$, $B_2 = (C/2)(-i + j)$, $B_3 = Ci$

$B = B_1 + B_2 + B_3$ $B = (\mu_0 I/4\pi L)(i - j)$

(c) Proceed as in part (a); note that B_3 is reversed $B_1 = -Cj$, $B_2 = (C/2)(i - j)$, $B_3 = -Ci$

$B = B_1 + B_2 + B_3$ $B = (\mu_0 I/4\pi L)(-i - 3j)$

41* •   A solenoid 2.7 m long has a radius of 0.85 cm and 600 turns. It carries a current I of 2.5 A. What is the approximate magnetic field B on the axis of the solenoid?

Use Equ. 29-9 $B = (4\pi\times10^{-7} \times600\times2.5/2.7)$ T $= 0.698$ mT

45* •   Ampère's law is valid

(a) when there is a high degree of symmetry.

(b) when there is no symmetry.

(c) when the current is constant.

(d) when the magnetic field is constant.

(e) in all of these situations if the current is continuous.

(e)

49* ••   A wire of radius 0.5 cm carries a current of 100 A that is uniformly distributed over its cross-sectional area. Find B (a) 0.1 cm from the center of the wire, (b) at the surface of the wire, and (c) at a point outside the wire 0.2 cm from the surface of the wire. (d) Sketch a graph of B versus the distance from the center of the wire.

We shall consider the general case first. Consider a wire of radius a carrying a current I of uniform current density.

1. For $r < a$, find I within area of radius $r < a$ $I(r) = Ir^2/a^2$

2. Use Equ. 29-15

3. For $r \geq a$, use Equ. 29-12

(a) Use the result for $r < a$; $I = 100$ A, $a = 0.5$ cm

(b) Use the result for $r \geq a$

(c) Use the result for $r > a$

(d) The graph of B versus r is shown.

$2\pi r B = \mu_0 I r^2 / a^2$; $B = \mu_0 I r / 2\pi a^2$

$B = \mu_0 I / 2\pi r$

$B = 8 \times 10^{-4}$ T

$B = 4 \times 10^{-3}$ T

$B = 2.86 \times 10^{-3}$ T

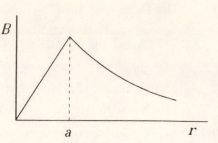

53* •• Figure 29-45 shows a solenoid carrying a current I with n turns per unit length. Apply Ampère's law to the rectangular curve shown to derive an expresion for B assuming that it is uniform inside the solenoid and zero outside it.

The number of turns enclosed within the rectangular area is na. Note that outside the solenoid $B = 0$ and within the solenoid on the paths along b, $B \cdot d\ell = 0$. Using Ampère's law, we have $aB = \mu_0 naI$ and $B = \mu_0 nI$.

57* • If the magnetic susceptibility is positive,

(a) paramagnetic effects or ferromagnetic effects must be greater than diamagnetic effects.

(b) diamagnetic effects must be greater than paramagnetic effects.

(c) diamagnetic effects must be greater than ferromagnetic effects.

(d) ferromagnetic effects must be greater than paramagnetic effects.

(e) paramagnetic effects must be greater than ferromagnetic effects.

(a)

61* • Repeat Problem 60 for a tungsten core.

1. Find B_{app} in the solenoid; use Equ. 29-9

2. $M = \chi_m B_{app}/\mu_0 = \chi_m nI$; $\chi_m = 6.8 \times 10^{-5}$ (see Table 29-1)

3. $B = B_{app}(1 + \chi_m)$

$B_{app} = 10.053$ mT

$M = 0.544$ A/m

$B = 10.054$ mT

65* •• An engineer intends to fill a solenoid with a mixture of oxygen and nitrogen at room temperature and 1 atmosphere pressure such that K_m is exactly 1. Assume that the magnetic dipole moments of the gas molecules are all aligned and that the susceptibility of a gas is proportional to the number density of its molecules. What should the ratio of the number densities of oxygen to nitrogen molecules be so that $K_m = 1$?

If $K_m = 1$, $\chi_m = 0$; so $n_O \chi_{mO} + n_N \chi_{mN} = 0$

$n_O/n_N = -\chi_{mN}/\chi_{mO} = 0.00239$

69* •• Nickel has a density of 8.7 g/cm³ and molecular mass of 58.7 g/mol. Its saturation magnetization is given by $\mu_0 M_s = 0.61$ T. Calculate the magnetic moment of a nickel atom in Bohr magnetons.

$\mu = M_s/(\mu_0 N_A \rho/\mathcal{M})$; number of $\mu_B = \mu/\mu_B$

$\mu = 0.587\ \mu_B$

73* •• Assume that the magnetic moment of an aluminum atom is 1 Bohr magneton. The density of aluminum is 2.7 g/cm³, and its molecular mass is 27 g/mol. (a) Calculate M_s and $\mu_0 M_s$ for aluminum. (b) Use the results of Problem 71 to calculate χ_m at $T = 300$ K. (c) Explain why the result for part (b) is larger than the value listed in Table 29-1.

(a) $M_s = (N_A \rho/\mathcal{M})\mu_B$;

$B_s = \mu_0 M_s$

$M_s = 6.02 \times 10^{28} \times 9.27 \times 10^{-24}$ A/m $= 5.58 \times 10^5$ A/m

$B_s = 0.701$ T

(b) $\chi_m = \mu_B\mu_0 M_s/3kT = \mu_B B_s/3kT$ $\chi_m = 9.27\times10^{-24}\times0.701/(3\times1.38\times10^{-23}\times300) = 5.23\times10^{-4}$

(c) In arriving at χ_m in part (b) we have neglected any diamagnetic effects.

77* • For annealed iron, the relative permeability K_m has its maximum value of about 5500 at $B_{app} = 1.57\times10^{-4}$ T. Find M and B when K_m is maximum.

$B = K_m B_{app}$; $M = (K_m - 1)B_{app}/\mu_0 \approx K_m B_{app}/\mu_0$ $B = 0.864$ T; $M = 6.87\times10^5$ A/m

---

81* •• When the current in Problem 80 is 0.2 A, the magnetic field is measured to be 1.58 T. (a) Neglecting end effects, what is B_{app}? (b) What is M? (c) What is the relative permeability K_m?

(a) Find $B_{app} = \mu_0 nI$ $B_{app} = 1.26\times10^{-3}$ T

(b) $M = (B - B_{app})/\mu_0$ $M = 1.26\times10^6$ A/m

(c) $K_m = B/B_{app}$ $K_m = 1.25\times10^3$

85* •• Find the magnetic field in the toroid of Problem 76 if the current in the wire is 0.2 A and soft iron, having a relative permeability of 500, is substituted for the paramagnetic core?

1. Find B_{app} using Equ. 29-9 $B_{app} = 1.508$ mT

2. $B = K_m B_{app}$ $B = 0.754$ T

89* • Can a particle have a magnetic moment and not have angular momentum?

No

93* • Find the magnetic field at point P in Figure 29-49.

B only due to semicircle; $B = \mu_0 I/4R$ $B = 2.36\times10^{-5}$ T

97* •• A loop of wire of length ℓ carries a current I. Compare the magnetic fields at the center of the loop when it is (a) a circle, (b) a square, and (c) an equilateral triangle. Which field is largest?

(a) Express B in terms of ℓ $B = \mu_0 I/2R = \pi\mu_0 I/\ell = 3.14\mu_0 I/\ell$

(b) Use Equ. 29-11; $R = \ell/8$; $\theta_1 = \theta_2 = 45°$ $B = [(\mu_0/4\pi)8I/\ell](2\sin\theta) = 8\sqrt{2}\mu_0 I/\pi\ell = 3.60\mu_0 I/\ell$

(c) Use Equ. 29-11; $R = (\ell/3)/(2\sqrt{3})$; $\theta_1 = \theta_2 = 60°$ $B = 27\mu_0 I/2\pi\ell = 4.30\mu_0 I/\ell$

The largest B is for the equilateral triangle

101* •• A closed circuit consists of two semicircles of radii 40 and 20 cm that are connected by straight segments as shown in Figure 29-54. A current of 3.0 A flows around this circuit in the clockwise direction. Find the magnetic field at point P.

Proceed as in Problem 93; $B = (\mu_0 I/4)(1/R_1 + 1/R_2)$ $B = 7.07\ \mu$T

105* •• Figure 29-55 shows a bar magnet suspended by a thin wire that provides a restoring torque $-\kappa\theta$. The magnet is 16 cm long, has a mass of 0.8 kg, a dipole moment of $\mu = 0.12$ A·m², and it is located in a region where a uniform magnetic field B can be established. When the external magnetic field is 0.2 T and the magnet is given a small angular displacement $\Delta\theta$, the bar magnet oscillates about its equilibrium position with a period of 0.500 s. Determine the constant κ and the period of this torsional pendulum when $B = 0$.

1. Write expression for ω^2 for $B = 0$ $-\kappa\theta = I(d^2\theta/dt^2)$; $\omega^2 = \kappa/I$, where $I = 1.71\times10^{-3}$ kg·m²

2. For $B \neq 0$, additional restoring τ is $\mu B\theta$, for $\theta \ll 1$ $\omega^2 = (\kappa + \mu B)/I = 4\pi^2/T^2$; $T^2 = 4\pi^2 I/(\kappa + \mu B)$

3. Solve for κ with $\mu = 0.12$ A·m^2, $B = 0.2$ T, $T = 0.5$ s $\kappa = 0.246$ N·m/rad

4. Find T for $B = 0$; $T = 2\pi\sqrt{I/\kappa}$ $T = 0.523$ s

---

109*••• The needle of a magnetic compass has a length of 3 cm, a radius of 0.85 mm, and a density of 7.96×10^3 kg/m^3. It is free to rotate in a horizontal plane, where the horizontal component of the earth's magnetic field is 0.6 G. When disturbed slightly, the compass executes simple harmonic motion about its midpoint with a frequency of 1.4 Hz. (*a*) What is the magnetic dipole moment of the needle? (*b*) What is the magnetization *M*? (*c*) What is the amperian current on the surface of the needle? (See Problem 106.)

(*a*) 1. Find $I = mL^2/12$; $m = \rho\pi r^2 L$; $I = \rho\pi r^2 L^3/12$ $I = 4.065 \times 10^{-8}$ kg·m^2

2. $f = (1/2\pi)\sqrt{\mu B/I}$; $\mu = 4\pi^2 f^2 I/B$ $\mu = 5.24 \times 10^{-2}$ A·m^2

(*b*) $M = \mu/V = \mu/\pi r^2 L$ $M = 7.70 \times 10^5$ A/m

(*c*) $I_{Amp} = M\Delta L$ $I_{Amp} = 2.31 \times 10^4$ A

113*••• An infinitely long, straight wire is bent as shown in Figure 29-56. The circular portion has a radius of 10 cm with its center a distance *r* from the straight part. Find *r* such that the magnetic field at the center of the circular portion is zero.

1. Let *R* be the radius of the loop; write B_{loop} $B_{loop} = \mu_0 I/2R$

2. Write B_{line} for infinite line $B_{line} = \mu_0 I/2\pi r$

3. Note that B_{line} is opposite to B_{loop}; find *r* for $B = 0$ $r = R/\pi = 3.18$ cm

---

117*••• In the Bohr model of the hydrogen atom, an electron in the ground state orbits a proton at a radius of 5.29×10^{-11} m. In a reference frame in which the orbiting electron is at rest, the proton circulates about the electron at a radius of 5.29×10^{-11} m with the same angular velocity as that of the electron in the reference frame in which the proton is at rest. Consequently, in the rest frame of the electron, the magnetic field due to the motion of the proton has the same magnitude as that calculated in Problem 5. The electron has an intrinsic magnetic moment of magnitude μ_B. Find the energy difference between the two possible orientations of the electron's intrinsic magnetic moment, either parallel or antiparallel to the magnetic field due to the apparent motion of the proton. (This energy difference is readily observed spectroscopically and is known as the *fine structure splitting*.)

$\Delta E = 2\mu_B B$; $B = 12.5$ T (see Problem 5) $\Delta E = 2 \times 9.27 \times 10^{-24} \times 12.5$ J $= 2.32 \times 10^{-22}$ J $= 1.45$ meV

121*•••• A very long, straight conductor with a circular cross section of radius *R* carries a current *I*. Inside the conductor, there is a cylindrical hole of radius *a* whose axis is parallel to the axis of the conductor a distance *b* from it (Figure 29-59). Let the *z* axis be the axis of the conductor, and let the axis of the hole be at $x = b$. Find the magnetic field *B* at the point (*a*) on the *x* axis at $x = 2R$, and (*b*) on the *y* axis at $y = 2R$. (*Hint*: Consider a uniform current distribution throughout the cylinder of radius *R* plus a current in the opposite direction in the hole.)

We apply the *Hint* and use the following notation. Let I_t be the total current that would flow into the paper. Let I_o be the current that would then have to flow out of the paper in the region of the hole so that the net current in the conductor is I_i. We assume uniform current density *J*.

(*a*) 1. Write I_i, I_o, and I_t in terms of *J* $I_i = J\pi(R^2 - a^2)$; $I_t = \pi R^2 J = \dfrac{I_i R^2}{R^2 - a^2}$; $I_o = \pi a^2 J = \dfrac{I_i a^2}{R^2 - a^2}$

2. Use Equ. 29-12 and right-hand rule to find *B* $B = \dfrac{\mu_0}{2\pi}\left(\dfrac{I_o}{2R - b} - \dfrac{I_t}{2R}\right)j = \dfrac{\mu_0 I_i}{2\pi(R^2 - a^2)}\left(\dfrac{a^2}{2R - b} - \dfrac{R}{2}\right)j$

(b) 1. Find $B_t = B$ due to I_t at $y = 2R$

$$B_t = \frac{\mu_0 I_t}{4\pi R} i$$

2. Find B_o at $y = 2R$ and its x and y components

$$B_o = \frac{\mu_0 I_o}{2\pi\sqrt{b^2 + 4R^2}}\, ; \; B_{ox} = -B_o \cos\theta = -B_o \frac{2R}{\sqrt{b^2 + 4R^2}}\, ;$$

$$B_{oy} = -B_o \frac{b}{\sqrt{b^2 + 4R^2}}$$

3. Find $B = B_t + B_o$

$$B = \frac{\mu_0 I_i}{\pi(R^2 - a^2)}\left[\left(\frac{R}{4} - \frac{a^2 R}{b^2 + 4R^2}\right) i - \left(\frac{a^2 b}{2(b^2 + 4R^2)}\right) j\right]$$

125*••• The current in a long cylindrical conductor of radius R is given by $I(r) = I_0(1 - e^{r/a})$. Derive expressions for the magnetic field for $r < R$ and for $r > R$.

For $r < R$, $B = (\mu_0 I_0/2\pi r)(1 - e^{-r/a})$; for $r > R$, $B = (\mu_0 I_0/2\pi r)(1 - e^{-R/a})$.

CHAPTER 30

Magnetic Induction

1* • A uniform magnetic field of magnitude 2000 G is parallel to the x axis. A square coil of side 5 cm has a single turn and makes an angle θ with the z axis as shown in Figure 30-28. Find the magnetic flux through the coil when (a) $\theta = 0°$, (b) $\theta = 30°$, (c) $\theta = 60°$, and (d) $\theta = 90°$.

(a), (b), (c), (d) $\phi_m = BA \cos \theta$

(a) $\phi_m = 2 \times 10^{-1} \times 25 \times 10^{-4}$ Wb $= 5 \times 10^{-4}$ Wb $= 0.5$ mWb;

(b) $\phi_m = 0.433$ mWb; (c) $\phi_m = 0.25$ mWb; (d) $\phi_m = 0$

5* • A uniform magnetic field B is perpendicular to the base of a hemisphere of radius R. Calculate the magnetic flux through the spherical surface of the hemisphere.

Note that ϕ_m through the base must also penetrate the spherical surface. Thus, $\phi_m = \pi R^2 B$.

9* •• A solenoid has n turns per unit length, radius R_1, and carries a current I. (a) A large circular loop of radius $R_2 > R_1$ and N turns encircles the solenoid at a point far away from the ends of the solenoid. Find the magnetic flux through the loop. (b) A small circular loop of N turns and radius $R_3 < R_1$ is completely inside the solenoid, far from its ends, with its axis parallel to that of the solenoid. Find the magnetic flux through this small loop.

(a) B for $R > R_1 = 0$; so $\phi_m = NBA = \mu_0 n I N \pi R_1^2$.

(b) Now $A = \pi R_3^2$, so $\phi_m = \mu_0 n I N \pi R_3^2$.

13* • A conducting loop lies in the plane of this page and carries a clockwise induced current. Which of the following statements could be true?

(a) A constant magnetic field is directed into the page.

(b) A constant magnetic field is directed out of the page.

(c) An increasing magnetic field is directed into the page.

(d) A decreasing magnetic field is directed into the page.

(e) A decreasing magnetic field is directed out of the page.

(d)

17* • The magnetic field in Problem 4 is steadily reduced to zero in 0.8 s. What is the magnitude of the emf induced in the coil of part (b)?

$\mathcal{E} = -d\phi_m/dt$ $\qquad\qquad |\mathcal{E}| = 0.015/0.8$ V $= 18.75$ mV

21* ••   A circular coil of 300 turns and radius 5.0 cm is connected to a current integrator. The total resistance of the circuit is 20 Ω. The plane of the coil is originally aligned perpendicular to the earth's magnetic field at some point. When the coil is rotated through 90°, the charge that passes through the current integrator is measured to be 9.4 μC. Calculate the magnitude of the earth's magnetic field at that point.

1. Relate the flux change to the charge

$\mathcal{E} = -\Delta\phi_m/\Delta t;\ I = \mathcal{E}/R;\ Q = I\Delta t = -\Delta\phi_m/R$

2. $\Delta\phi_m = NBA;\ B = QR/NA = QR/N\pi r^2$

$B = 79.8\ \mu$T

25* •   Give the direction of the induced current in the circuit on the right in Figure 30-32 when the resistance in the circuit on the left is suddenly (*a*) increased and (*b*) decreased.

Note that when R is constant, B in the loop to the right points out of the paper.

(*a*) If R increases, I decreases and so does B. By Lenz's law, the induced current is counterclockwise.

(*b*) Conversely, if R decreases, the induced current is clockwise.

29* •   A rod 30 cm long moves at 8 m/s in a plane perpendicular to a magnetic field of 500 G. The velocity of the rod is perpendicular to its length. Find (*a*) the magnetic force on an electron in the rod, (*b*) the electrostatic field E in the rod, and (*c*) the potential difference V between the ends of the rod.

(*a*) $\mathbf{F} = q\mathbf{v} \times \mathbf{B}$

$F = 1.6 \times 10^{-19} \times 8 \times 5 \times 10^{-2}$ N $= 6.4 \times 10^{-20}$ N

(*b*) $\mathbf{E} = \mathbf{v} \times \mathbf{B}$

$E = 0.4$ V/m

(*c*) $V = E\ell$

$V = 0.12$ V

33* ••   A 10-cm by 5-cm rectangular loop with resistance 2.5 Ω is pulled through a region of uniform magnetic field $B = 1.7$ T (Figure 30-36) with constant speed $v = 2.4$ cm/s. The front of the loop enters the region of the magnetic field at time $t = 0$. (*a*) Find and graph the flux through the loop as a function of time. (*b*) Find and graph the induced emf and the current in the loop as functions of time. Neglect any self-inductance of the loop and extend your graphs from $t = 0$ to $t = 16$ s.

(*a*) For $0 < t < 4.17$ s, $\phi_m = 0.05Bvt = 2.04 \times 10^{-3}t$;

for 4.17 s $< t <$ 8.33 s, $\phi_m = 8.51$ mWb;

for 8.33 s $< t <$ 12.5 s, $\phi_m = 8.51 \times 10^{-3}(t - 8.33)$;

for $t > 12.5$ s, $\phi_m = 0$

The graph of $\phi_m(t)$ is shown on the right. Here ϕ_m is in mWb, t in s.

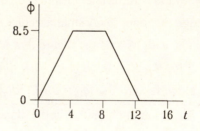

(*b*) $\mathcal{E} = -(d\phi_m/dt)$; for $0 < t < 4.17$ s, $\mathcal{E} = -2.04$ mV

for 4.17 s $< t <$ 8.33 s, $\mathcal{E} = 0$;

for 8.33 s $< t <$ 12.5 s, $\mathcal{E} = 2.04$ mV;

for $t > 12.5$ s, $\mathcal{E} = 0$.

The graph of $\mathcal{E}(t)$ is shown on the right. Here $\mathcal{E}$ is in mV and t in s.

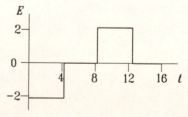

The graph of the current is identical to that of $\mathcal{E}$ with the ordinate scale changed to $I_{max} = 0.8$ mA.

37* ••   Find the total distance traveled by the rod in Example 30-8.

Note that $v = dx/dt = v_0 e^{-Ct}$, where $C = B^2\ell^2/mR$. Integrating the differential equation gives $x = \int_0^\infty v_0 e^{-Ct}dt = v_0/C$. The distance traveled is $v_0 mR/B^2\ell^2$.

41★ ••   A wire lies along the z axis and carries current $I = 20$ A in the positive z direction. A small conducting sphere of radius $R = 2$ cm is initially at rest on the y axis at a distance $h = 45$ m above the wire. The sphere is dropped at time $t = 0$. (*a*) What is the electric field at the center of the sphere at $t = 3$ s? Assume that the only magnetic field is that produced by the wire. (*b*) What is the voltage across the sphere at $t = 3$ s?

(*a*) 1. Find $B(y)$; use Equ. 29-12 $\qquad\qquad$ $B = 20\mu_0/2\pi y\, i = (4\times10^{-6}/y)$ T i

$\qquad$ 2. Find the velocity of the sphere at $t = 3$ s $\qquad$ $v = -gt\, j = -29.4$ m/s j

$\qquad$ 3. Find y at $t = 3$ s; $y = h - \frac{1}{2}gt^2$ and B $\qquad$ $y = 0.855$ m; $B = 1.17\times10^{-6}$ T i

$\qquad$ 4. Find $E = v\times B$ $\qquad\qquad\qquad\qquad\qquad$ $E = 0.138$ mV/m k

(*b*) $V = 2RE$ $\qquad\qquad\qquad\qquad\qquad\qquad\qquad$ $V = 5.50\ \mu$V

45★ •••   A rod of length ℓ is perpendicular to a long wire carrying current I, as shown in Figure 30-40. The near end of the rod is a distance d away from the wire. The rod moves with a speed v in the direction of the current I. (*a*) Show that the potential difference between the ends of the rod is given by

$$V = \frac{\mu_0 I}{2\pi}v\,\ln\frac{d+\ell}{d}$$

(*b*) Use Faraday's law to obtain this result by considering the flux through a rectangular area $A = \ell vt$ swept out by the rod.

(*a*) Consider a small segment of the rod of length dx. The induced field is $dE = B(x)v\,dx$, where $B(x) = \mu_0 I/2\pi x$.

Thus, $V = \displaystyle\int_d^{d+\ell} E\,dx = \int_d^{d+\ell}\frac{\mu_0 Iv}{2\pi x} = \frac{\mu_0 Iv}{2\pi}\ln\left(\frac{d+\ell}{d}\right)$.

(*b*) Again, consider a segment dx. $d\phi_m = B\,dA = B\,vt\,dx = (\mu_0 Iv/2\pi x)t\,dx$. Except for the factor t, the integrand is the same as before; hence

$$\phi_m = \frac{\mu_0 Ivt}{2\pi}\ln\left(\frac{d+\ell}{d}\right).$$ The induced emf is $\mathcal{E} = d\phi_m/dt = \dfrac{\mu_0 Iv}{2\pi}\ln\left(\dfrac{d+\ell}{d}\right)$.

49★ •   A coil with a self-inductance of 8.0 H carries a current of 3 A that is changing at a rate of 200 A/s. Find (*a*) the magnetic flux through the coil and (*b*) the induced emf in the coil.

(*a*) $\phi_m = LI$; $I = (3 + 200t)$ A $\qquad\qquad$ $\phi_m = 24 + 1600t$ Wb

(*b*) $\mathcal{E} = -L(dI/dt)$ $\qquad\qquad\qquad\qquad$ $\mathcal{E} = -1600$ V

53★ ••   A long, insulated wire with a resistance of 18 Ω/m is to be used to construct a resistor. First, the wire is bent in half, and then the doubled wire is wound in a cylindrical form as shown in Figure 30-42. The diameter of the cylindrical form is 2 cm, its length is 25 cm, and the total length of wire is 9 m. Find the resistance and inductance of this wire-wound resistor.

Note that the current in the two parts of the wire is in opposite directions. Consequently, the total flux in the coil is zero, and the self inductance is also zero. $L = 0$. The length of the wire is 9 m, so its resistance is 9×18 $\Omega = 162$ Ω.

57★ •   If the current through an inductor were doubled, the energy stored in the inductor would be (*a*) the same. (*b*) doubled. (*c*) quadrupled. (*d*) halved. (*e*) quartered.

(*c*)

61* ••   A solenoid of 2000 turns, area 4 cm², and length 30 cm carries a current of 4.0 A. (*a*) Calculate the magnetic energy stored in the solenoid from $\frac{1}{2}LI^2$. (*b*) Divide your answer in part (*a*) by the volume of the solenoid to find the magnetic energy per unit volume in the solenoid. (*c*) Find *B* in the solenoid. (*d*) Compute the magnetic energy density from $u_m = B^2/2\mu_0$, and compare your answer with your result for part (*b*).

(*a*) Use Equ. 30-9 to find *L*; $U_m = \frac{1}{2}LI^2$ $L = 6.70$ mH; $U_m = 53.6$ mJ

(*b*) $u_m = U_m/V = U_m/A\ell$ $u_m = 447$ J/m³

(*c*) $B = \mu_0 nI = \mu_0 NI/\ell$ $B = 33.5$ mT

(*d*) $u_m = B^2/2\mu_0$ $u_m = 447$ J/m³ as in part (*b*)

65* •   The current in a coil with a self-inductance of 1 mH is 2.0 A at *t* = 0, when the coil is shorted through a resistor. The total resistance of the coil plus the resistor is 10.0 Ω. Find the current after (*a*) 0.5 ms and (*b*) 10 ms.

(*a*), (*b*) $I(t) = I_0 e^{-(R/L)t}$; $R/L = 10^4$ s⁻¹   (*a*) $I(0.5$ ms$) = 13.5$ mA (*b*) $I(10$ ms$) = 7.44 \times 10^{-44}$ A ≈ 0

69* ••   How many time constants must elapse before the current in an *RL* circuit that is initially zero reaches (*a*) 90%, (*b*) 99%, and (*c*) 99.9% of its final value?

(*a*), (*b*), (*c*) $1 - e^{-t/\tau} = x$; $t/\tau = \ln[1/(1 - x)]$ (*a*) $x = 0.9$, $t/\tau = 2.30$; (*b*) $x = 0.99$, $t/\tau = 4.61$;

        (*c*) $x = 0.999$, $t/\tau = 6.91$

73* ••   Compute the initial slope dI/dt at *t* = 0 from Equation 30-24, and show that if the current decreased steadily at this rate, it would be zero after one time constant.

$I = I_0 e^{-t/\tau}$; $dI/dt = -(1/\tau)I_0 e^{-t/\tau}$. At *t* = 0, $dI/dt = -I_0/\tau$. If *I* is a linear function of *t* then $I = I_0(1 - t/\tau)$ and $I = 0$ at *t* = *τ*.

77* ••   For the circuit of Example 30-11, find the time at which the power dissipation in the resistor equals the rate at which magnetic energy is stored in the inductor.

1. Find the rate of energy storage in *L* $dU_L/dt = d(\frac{1}{2}LI^2)/dt = LI(dI/dt)$

2. Set $I^2R = LI(dI/dt)$ $1/\tau = (1/I)(dI/dt) = e^{-t/\tau}/[\tau(1 - e^{-t/\tau})]$

3. Solve for t/τ and find *t* for $\tau = 333$ *μ*s $t/\tau = \ln(2)$; $t = 231$ *μ*s

81* •   A bar magnet is dropped inside a long vertical tube. If the tube is made of metal, the magnet quickly approaches a terminal speed, but if the tube is made of cardboard, it does not. Explain.

The time varying magnetic field of the magnet sets up eddy currents in the metal tube. The eddy currents establish a magnetic field with a magnetic moment opposite to that of the moving magnet; thus the magnet is slowed down. If the tube is made of a nonconducting material, there are no eddy currents.

85* ••   Figure 30-48 shows an ac generator. It consists of a rectangular loop of dimensions *a* and *b* with *N* turns connected to slip rings. The loop rotates with an angular velocity *ω* in a uniform magnetic field *B*. (*a*) Show that the potential difference between the two slip rings is $E = NBab\omega \sin wt$. (*b*) If *a* = 1.0 cm, *b* = 2.0 cm, *N* = 1000, and *B* = 2 T, at what angular frequency *ω* must the coil rotate to generate an emf whose maximum value is 110 V?

(*a*) Find $\phi_m(t)$ and $\mathcal{E} = -d\phi_m/dt$ $\phi_m(t) = NBA \cos \omega t$; $\mathcal{E} = NBab\omega \sin \omega t$

(*b*) $\mathcal{E}_{max}$ for $\sin \omega t = 1$; solve for *ω* $\omega = 110/(1000 \times 2 \times 10^{-4} \times 2) = 275$ rad/s

89*••   Suppose the coil of Problem 88 is rotated about its vertical centerline at constant angular velocity of 2 rad/s.
Find the induced current as a function of time.

From Problem 85, $I = \mathcal{E}/R = (NBA\omega \sin \omega t)/R$ $I(t) = (80\times0.0375\times1.4\times2/24) \sin 2t$ A$= 0.35 \sin 2t$ A

93* ••   A long solenoid has n turns per unit length and carries a current given by $I = I_0 \sin wt$. The solenoid has a
circular cross section of radius R. Find the induced electric field at a radius r from the axis of the solenoid for
(a) $r < R$ and (b) $r > R$.

(a) The field within the solenoid is $B = \mu_0 nI$. The flux through an area πr^2 for $r < R$ is then $\phi_m = \pi r^2 \mu_0 nI$. We now
apply Equ. 30-5 and obtain $2\pi rE = -d\phi_m/dt$. Solving for E we find $E = -(\mu_0 nrI_0\omega/2) \cos \omega t$.

(b) Proceed as in part (a) with $\phi_m = \pi R^2 \mu_0 nI$. One obtains $E = -(\mu_0 nR^2 I_0\omega/2r) \cos \omega t$.

97* •••   Figure 30-51 shows a rectangular loop of wire, 0.30 m wide and 1.50 m long, in the vertical plane and
perpendicular to a uniform magnetic field $B = 0.40$ T, directed inward as shown. The portion of the loop not in the
magnetic field is 0.10 m long. The resistance of the loop is 0.20 Ω and its mass is 0.50 kg. The loop is released
from rest at $t = 0$. (a) What is the magnitude and direction of the induced current when the loop has a downward
velocity v? (b) What is the force that acts on the loop as a result of this current? (c) What is the net force acting on
the loop? (d) Write the equation of motion of the loop. (e) Obtain an expression for the velocity of the loop as a
function of time. (f) Integrate the expression obtained in part (e) to find the displacement y as a function of time. (g)
From the result obtained in part (f) find t for $y = 1.40$ m, i.e., the time when the loop leaves the region of magnetic
field. (h) Find the velocity of the loop at that instant. (i) What would be the velocity of the loop after it has dropped
1.40 m if $B = 0$?

(a) $\mathcal{E} = Bv\ell$; $I = \mathcal{E}/R$; use Lenz's law $I = (0.4\times0.3/0.2)v$ A $= 0.6v$ A, clockwise

(b) $F_v = BI\ell$ directed up $F_v = (0.4\times0.3\times0.6)v$ A $= 0.072v$ N, directed up

(c) $F_{net} = mg - F_v$ $F_{net} = (0.5g - 0.072v)$ N, directed down

(d) $F_{net} = ma = m(dv/dt)$ $0.5g - 0.072v = 0.5(dv/dt)$

(e) See Problem 5-88 $v(t) = 68.1(1 - e^{-0.144t})$ m/s

(f) $y(t) = \int v(t)\, dt$ $y(t) = 68.1[t + 6.94(e^{-0.144t} - 1)]$ m

(g) Find t for $y = 1.40$ m by trial and error $t = 0.538$ s

(h) Find $v(0.538$ s$)$ from part (e) $v(0.538$ s$) = 5.08$ m/s

(i) $v = \sqrt{2gh}$ $v = 5.24$ m/s

CHAPTER 31

Alternating-Current Circuits

Note: Unless otherwise indicated, the symbols I, V, $\mathcal{E}$, and P denote the rms values of I, V, and $\mathcal{E}$ and the average power.

1★ • A 200-turn coil has an area of 4 cm² and rotates in a magnetic field of 0.5 T. (*a*) What frequency will generate a maximum emf of 10 V? (*b*) If the coil rotates at 60 Hz, what is the maximum emf?

(*a*) $\mathcal{E} = NBA\omega \cos \omega t$ (see Problem 30-85)

$\omega = \mathcal{E}_{max}/NBA = 250 \text{ s}^{-1}; f = \omega/2\pi = 39.8 \text{ Hz}$

(*b*) $\mathcal{E}_{max} = NBA\omega = 2\pi NBAf$

$\mathcal{E}_{max} = 15.1 \text{ V}$

5★ • As the frequency in the simple ac circuit in Figure 31-26 increases, the rms current through the resistor (*a*) increases. (*b*) does not change. (*c*) may increase or decrease depending on the magnitude of the original frequency. (*d*) may increase or decrease depending on the magnitude of the resistance. (*e*) decreases.

(*b*)

9★ • A circuit breaker is rated for a current of 15 A rms at a voltage of 120 V rms. (*a*) What is the largest value of I_{max} that the breaker can carry? (*b*) What average power can be supplied by this circuit?

(*a*) $I_{max} = \sqrt{2}I_{rms}$

$I_{max} = 21.2 \text{ A}$

(*b*) $P = I_{rms}V_{rms}$

$P = 1.8 \text{ kW}$

13★ •• In a circuit consisting of a generator and an inductor, are there any times when the inductor absorbs power from the generator? Are there any times when the inductor supplies power to the generator?

Yes, Yes

17★ •• At what frequency would the reactance of a 10.0-μF capacitor equal that of a 1.0-mH inductor?

$f = (1/2\pi)(1/\sqrt{LC})$

$f = 1.59 \text{ kHz}$

21★ •• Draw the resultant phasor diagram for a series RLC circuit when $V_L < V_C$. Show on your diagram that the emf will lag the current by the phase angle δ given by

$$\tan \delta = \frac{V_C - V_L}{V_R}$$

The phasor diagram is shown at the right. The voltages V_R, V_L, and V_C are indicated as well as the resultant voltage E. The current is in phase with V_R and its phasor is shown by the dashed arrow. The voltage E lags the current by the angle δ where $\delta = \tan^{-1}[(V_C - V_L)/V_R]$.

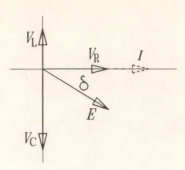

25* • Show from the definitions of the henry and the farad that $1/\sqrt{LC}$ has the unit s^{-1}.

The dimension of C is [Q]/[V]. From $V = L(dI/dt)$, and [I] = [Q]/[T] it follows that [L] = [V][T]^2/[Q].

Thus [L][C] = [T]^2 and $1/\sqrt{LC}$ has the dimension of of [T]$^{-1}$, i.e., units of s^{-1}.

29* • A coil can be considered to be a resistance and an inductance in series. Assume that $R = 100$ Ω and $L = 0.4$ H. The coil is connected across a 120-V-rms, 60-Hz line. Find (*a*) the power factor, (*b*) the rms current, and (*c*) the average power supplied.

(*a*) $X = X_L = \omega L$; $Z = \sqrt{X^2 + R^2}$; pf = R/Z $X_L = 150.8$ Ω; $Z = 181$ Ω; power factor = 0.552

(*b*) $I = \mathcal{E}/Z$ $I = 120/181$ A = 0.663 A

(*c*) $P = I^2R$ $P = 44.0$ W

33* •• A coil with resistance and inductance is connected to a 120-V-rms, 60-Hz line. The average power supplied to the coil is 60 W, and the rms current is 1.5 A. Find (*a*) the power factor, (*b*) the resistance of the coil, and (*c*) the inductance of the coil. (*d*) Does the current lag or lead the voltage? What is the phase angle δ?

(*a*) $P = \mathcal{E}I \times$pf   pf = cos δ = 60/180 = 0.333; $\delta = 70.5°$

(*b*) $R = P/I^2$ $R = 60/2.25$ Ω = 26.7 Ω

(*c*) $X_L = R \tan \delta = \omega L$; $L = (R \tan \delta)/\omega$ $L = 0.2$ H

(*d*) The circuit is inductive I lags $\mathcal{E}$; $\delta = 70.5°$

37* ••• Figure 31-31 shows a load resistor $R_L = 20$ Ω connected to a high-pass filter consisting of an inductor $L = 3.2$ mH and a resistor $R = 4$ Ω. The input voltage is $\mathcal{E} = (100$ V) cos $(2\pi ft)$. Find the rms currents in R, L, and R_L if (*a*) $f = 500$ Hz and (*b*) $f = 2000$ Hz. (*c*) What fraction of the total power delivered by the voltage source is dissipated in the load resistor if the frequency is 500 Hz and if the frequency is 2000 Hz?

We shall do this problem for the general case and then substitute numerical values.

1. Find the resistive and inductive components of $R_p = R_L \omega^2 L^2/(R_L^2 + \omega^2 L^2)$; $X_p = \omega L R_L^2/(R_L^2 + \omega^2 L^2)$
   $Z_p = Z$ of the parallel combination of L and R_L $Z_p = R_L \omega L/\sqrt{R_L^2 + \omega^2 L^2}$

2. Find $I = I_R$ in terms of other parameters $I = \mathcal{E}/\sqrt{(R + R_p)^2 + (X_p)^2} = I_R$

3. Write V_p, the voltage across Z_p $V_p = IZ_p$

4. Write the currents in L and R_L $I_L = IZ_p/\omega L$; $I_{R_L} = IZ_p/R_L$

5. Write the power dissipated in R and in R_L $P_R = I^2R$; $P_L = I_{R_L}^2 R_L$; $P_{tot} = P_R + P_L$

(*a*) 1. For $f = 500$ Hz, find R_p, X_p, and Z_p $R_p = 4.03$ Ω, $X_p = 8.02$ Ω, $Z_p = 8.98$ Ω

     2. Find $I = I_R$ $I = I_R = 100/(8.03^2 + 8.02^2)^{1/2}$ A = 8.81 A

     3. Find I_L and I_{R_L} $I_L = 8.81 \times 8.98/10.05$ A = 7.87 A; $I_{R_L} = 3.96$ A

(b) 1. For $f = 2000$ Hz, find R_p, X_p, and Z_p $\qquad$ $R_p = 16.0\ \Omega$, $X_p = 7.98\ \Omega$, $Z_p = 17.9\ \Omega$

$\quad$ 2. Find $I = I_R$ $\qquad$ $I = I_R = 100/(20^2 + 7.98^2)^{\frac{1}{2}}$ A $= 4.64$ A

$\quad$ 3. Find I_L and I_{R_L} $\qquad$ $I_L = 4.64 \times 17.9/40.2$ A $= 2.07$ A; $I_{R_L} = 4.16$ A

Note: As $f \to \infty$, $I_R = I_{R_L} = 5.00$ A

(c) 1. For $f = 500$ Hz, find P_R, P_L, P_{tot}, and P_L/P_{tot} $\qquad$ $P_R = 310$ W, $P_L = 314$ W; $P_{tot} = 624$ W; $P_L/P_{tot} = 50.3\%$

$\quad$ 2. Repeat above for $f = 2000$ Hz $\qquad$ $P_R = 86.1$ W, $P_L = 346$ W; $P_{tot} = 432$ W; $P_L/P_{tot} = 80.0\%$

41* •• A coil draws 15 A when connected to a 220-V 60-Hz ac line. When it is in series with a 4-Ω resistor and the combination is connected to a 100-V battery, the battery current after a long time is observed to be 10 A. (a) What is the resistance in the coil? (b) What is the inductance of the coil?

(a) For $t \to \infty$, $I_B = \mathcal{E}_B/(R_L + 4.0)$; solve for R_L $\qquad$ $R_L = 6.0\ \Omega$

(b) $Z = \mathcal{E}/I$; $L = \sqrt{Z^2 - R_L^2}/\omega$; $\omega = 377$ s^{-1} $\qquad$ $Z = 14.7\ \Omega$, $L = 35.5$ mH

45* ••• One method for measuring the compressibility of a dielectric material uses an LC circuit with a parallel-plate capacitor. The dielectric is inserted between the plates and the change in resonance frequency is determined as the capacitor plates are subjected to a compressive stress. In such an arrangement, the resonance frequency is 120 MHz when a dielectric of thickness 0.1 cm and dielectric constant $\kappa = 6.8$ is placed between the capacitor plates. Under a compressive stress of 800 atm, the resonance frequency decreases to 116 MHz. Find Young's modulus of the dielectric material.

We shall do this problem for the general case and then substitute numerical values. Let t be the initial thickness of dielectric. Then $C_0 = \kappa\epsilon_0 A/t$ and $C_p = \kappa\epsilon_0 A/(t - \Delta t) = C_0/(1 - \Delta t/t)$ is the capacitance under compression. We have $\omega_0 = 1/(C_0 L)^{\frac{1}{2}}$ and $\omega_p = 1/(C_p L)^{\frac{1}{2}}$. $\omega_p/\omega_0 = (1 - \Delta t/t)^{\frac{1}{2}} \approx 1 - \Delta t/2t$ since $\omega_p/\omega_0 = 1 - \epsilon$, where $\epsilon \ll 1$. From the definition of Young's modulus we have $Y = \text{stress}/(\Delta t/t)$.

1. Find $\Delta t/t$ $\qquad$ $\Delta t/t = 2 \times 4/120 = 0.0667$

2. Determine Y; stress $= 808 \times 10^5$ N/m^2 $\qquad$ $Y = 808 \times 10^5/0.0667 = 1.21 \times 10^9$ N/m^2

49* • Are there any disadvantages to having a radio tuning circuit with an extremely large Q factor?

Yes; the bandwidth must be wide enough to accommodate the modulation frequency.

53* • An ac generator with a maximum emf of 20 V is connected in series with a 20-μF capacitor and an 80-Ω resistor. There is no inductance in the circuit. Find (a) the power factor, (b) the rms current, and (c) the average power if the angular frequency of the generator is 400 rad/s.

(a) $Z = \sqrt{R^2 + 1/\omega^2 C^2}$; power factor $= R/Z$ $\qquad$ $Z = 148\ \Omega$; power factor $= 0.539$

(b) $I = \mathcal{E}/Z$; $\mathcal{E} = \mathcal{E}_{max}/\sqrt{2}$ $\qquad$ $I = 14.1/148$ A $= 0.0956$ A

(c) $P = I^2 R$ $\qquad$ $P = 0.731$ W

57* •• Find the power factor and the phase angle δ for the circuit in Problem 55 when the generator frequency is (a) 900 Hz, (b) 1.1 kHz, and (c) 1.3 kHz.

(a) Find X and Z; $X = \omega L - 1/\omega C$; $\omega = 5655$ s^{-1} $\qquad$ $X = -31.9\ \Omega$; $Z = 32.3\ \Omega$; $\cos \delta = 0.155$; $\delta = -81.1°$

(b) Repeat part (a) with $\omega = 6912$ s^{-1} $\qquad$ $X = -3.2\ \Omega$; $Z = 5.94\ \Omega$; $\cos \delta = 0.842$; $\delta = -32.6°$

(c) Repeat part (a) with $\omega = 8168$ s^{-1} $\qquad$ $X = 20.5\ \Omega$; $Z = 21.1\ \Omega$; $\cos \delta = 0.237$; $\delta = 76.3°$

61* ••   A 0.25-H inductor and a capacitor C are connected in series with a 60-Hz ac generator. An ac voltmeter is used to measure the rms voltages across the inductor and capacitor separately. The rms voltage across the capacitor is 75 V and that across the inductor is 50 V. (*a*) Find the capacitance C and the rms current in the circuit. (*b*) What would be the measured rms voltage across both the capacitor and inductor together?

(*a*) 1. Find $I = V_L/\omega L$ $I = 50/(377 \times 0.25)$ A $= 0.5305$ A

   2. $I/\omega C = V_C$; find C $C = 0.5305/(75 \times 377)$ F $= 18.8\ \mu$F

(*b*) Since $R = 0$, $V = |V_L - V_C|$ $V = 25$ V

65* ••   A variable-frequency ac generator is connected to a series RLC circuit for which $R = 1$ kΩ, $L = 50$ mH, and $C = 2.5\ \mu$F. (*a*) What is the resonance frequency of the circuit? (*b*) What is the Q value? (*c*) At what frequencies is the value of the average power delivered by the generator half of its maximum value?

(*a*) $f_0 = 1/2\pi\sqrt{LC}$ $f_0 = 450$ Hz

(*b*) $Q = \omega_0 L/R = \dfrac{\sqrt{L/C}}{R}$ $Q = 0.141$

(*c*) When $\omega = \omega_0$, P is a maximum: $P = \mathcal{E}^2/R$. When $\omega \neq \omega_0$, P is given by Equ. 31-58. Set Equ. 31-58 equal to $\mathcal{E}^2/2R$. This gives $R^2\omega^2 = L^2(\omega^2 - \omega_0^2)^2$, or $L^2\omega^4 - (2L^2\omega_0^2 + R^2)\omega^2 + L^2\omega_0^2 = 0$. The quadratic equation has the solution

$$\omega^2 = \frac{(2L^2\omega_0^2 + R^2) \pm R\sqrt{4L^2\omega_0^2 + R^2}}{2L^2}.$$ Substituting appropriate numerical values one obtains

$\omega^2 = 4.158 \times 10^8$ s^{-2} and $\omega^2 = 1.537 \times 10^5$ s^{-2}. The corresponding (positive) frequencies are 3.25 kHz and 62.4 Hz.

69* ••   A series RLC circuit with $R = 400\ \Omega$, $L = 0.35$ H, and $C = 5\ \mu$F is driven by a generator of variable frequency f. (*a*) What is the resonance frequency f_0? Find f and f/f_0 when the phase angle δ is (*b*) $60°$, and (*c*) $-60°$.

(*a*) $f_0 = 1/2\pi\sqrt{LC}$ $f_0 = 120$ Hz

(*b*) From Equ. 31-51, $R \tan \delta = \omega L - 1/\omega C$; solve for $0.35\omega^2 - 693\omega - 2 \times 10^5 = 0$; $\omega = 2.24 \times 10^3$ s^{-2}; $f = 356$ Hz

   ω with $\delta = +60°$ and $\delta = -60°$.   and for $\delta = -60°$, $f = 40.7$ Hz.

   List f/f_0 for the two cases $\delta = 60°$, $f/f_0 = 2.96$; $\delta = -60°$, $f/f_0 = 0.338 = 1/2.96$

73* ••   In a series RLC circuit, $X_C = 16\ \Omega$ and $X_L = 4\ \Omega$ at some frequency. The resonance frequency is $\omega_0 = 10^4$ rad/s. (*a*) Find L and C. If $R = 5\ \Omega$ and $\mathcal{E}_{max} = 26$ V, find (*b*) the Q factor and (*c*) the maximum current.

(*a*) 1. Write the expressions for the known data $LC = 10^{-8}$ s^2; $\omega L = 4\ \Omega$, $1/\omega C = 16\ \Omega$; $L/C = 64$ H/F

   2. Solve for C and L $C = 12.5\ \mu$F; $L = 0.8$ mH

(*b*) $Q = \dfrac{\sqrt{L/C}}{R}$ $Q = 1.6$; $I_{max} = 5.2$ A

(*c*) $I_{max} = \mathcal{E}_{max}/Z$ $Z = \sqrt{25 + 144}\ \Omega = 13\ \Omega$; $I_{max} = 2.0$ A

77* ••   A resistor and a capacitor are connected in parallel across a sinusoidal emf $\mathcal{E} = \mathcal{E}_{max} \cos \omega t$ as shown in Figure 31-39. (*a*) Show that the current in the resistor is $I_R = (\mathcal{E}_{max}/R) \cos \omega t$. (*b*) Show that the current in the capacitor branch is $I_C = (\mathcal{E}_{max}/X_C) \cos (\omega t + 90°)$. (*c*) Show that the total current is given by $I = I_R + I_C = I_{max} \cos (\omega t + \delta)$, where $\tan \delta = R/X_C$ and $I_{max} = \mathcal{E}_{max}/Z$ with $Z^{-2} = R^{-2} + X_C^{-2}$.

(*a*) From Ohm's law, $I_R(t) = V(t)/R$. Here $V(t) = \mathcal{E}(t) = \mathcal{E}_{max} \cos \omega t$, so $I_R(t) = (\mathcal{E}_{max}/R) \cos \omega t$.

(*b*) For the capacitor, $V_C(t) = \mathcal{E}(t)$ and $V_C(t) = q(t)/C$; consequently, $d\mathcal{E}/dt = d(q(t)/C)/dt = I_C(t)/C$.

$d\mathcal{E}/dt = -\mathcal{E}_{max}\,\omega \sin \omega t = \mathcal{E}_{max}\,\omega \cos(\omega t + 90°)$. Hence, $I_C(t) = (\mathcal{E}_{max}/X_C)\cos(\omega t + 90°)$, where $X_C = 1/\omega C$.

(c) From Kirchhoff's law, $I = I_R + I_C = \mathcal{E}_{max}[(1/R)\cos \omega t - (1/X_C)\sin \omega t]$. If we write $I = I_{max}\cos(\omega t + \delta)$ and use the trigonometric identity for $\cos(\alpha + \beta) = \cos \alpha \cos \beta - \sin \alpha \sin \beta$, $I = I_{max}(\cos \omega t \cos \delta - \sin \omega t \sin \delta)$. Comparing this expression with I as given in terms of R and X_C, we see that $\tan \delta = R/X_C$. If we now define $Z = \sqrt{R^2 + X_C^2}$, then $\sin \delta = R/Z$ and $\cos \delta = X_C/Z$, and $I_{max} = \mathcal{E}_{max}/Z$.

81* ••   For the circuit in Figure 31-23, derive an expression for the Q of the circuit, assuming the resonance is sharp. Q is defined as $\omega_0/\Delta\omega$, where $\Delta\omega$ is the width of the resonance at half maximum. The currents in the three circuit elements are $I_C = V/X_C = \omega CV$, $I_L = V/\omega L$, and $I_R = V/R$, with I_C leading V and I_L lagging V by 90°. The total current is therefore $I = V\sqrt{(1/R)^2 + (\omega C - 1/\omega L)^2} = (V/R)\sqrt{1 + R^2(\omega C - 1/\omega L)^2}$. At resonance, the reactive term is zero and $I_0 = V/R$. The stored energy per cycle will be at half-maximum when $R(\omega C - 1/\omega L) = \pm 1$. This gives quadratic equations for ω with two solutions ω_+ and ω_- whose difference is $\Delta\omega = 1/RC$. Using $\omega_0 = 1/\sqrt{LC}$ and $Q = \omega_0/\Delta\omega$ one obtains $Q = R\sqrt{C/L}$.

85* ••   Repeat Problem 84 with the 100-Ω resistor replaced by a 40-Ω resistor.

(a) $f_0 = (1/2\pi)\sqrt{1/LC}$ $\qquad\qquad\qquad\qquad$ $f_0 = 13.26$ kHz

(b) At $f = f_0$, $I = \mathcal{E}/R$; $V_L = \omega_0 LI$; $V_C = V_L$ $\qquad$ $I = 0.5$ A; $V_L = V_C = 1.50$ kV

(c) $\tfrac{1}{2}\Delta f = f_0/2Q = f_0 R/2\omega_0 L$; find $f_0 + \tfrac{1}{2}\Delta f$ $\quad$ $f_0 + \tfrac{1}{2}\Delta f = f_0(1 + R/2\omega_0 L) = 13.26(1 + 0.0067)$ kHz

$\qquad X = R$, so $I = I_0/\sqrt{2}$; $V_L = IX_L$, $V_C = IX_C$ $\quad$ $I = 0.354$ A; $V_L = 1068$ V, $V_C = 1055$ V

89* •••   An ac generator is in series with a capacitor and an inductor in a circuit with negligible resistance. (a) Show that the charge on the capacitor obeys the equation

$$L\frac{d^2Q}{dt^2} + \frac{Q}{C} = \mathcal{E}_{max}\cos \omega t$$

(b) Show by direct substitution that this equation is satisfied by $Q = Q_{max}\cos \omega t$ if

$\qquad Q_{max} = -\mathcal{E}_{max}/[L(\omega^2 - \omega_0^2)]$

(c) Show that the current can be written as $I = I_{max}\cos(\omega t - \delta)$, where

$\qquad I_{max} = \omega \mathcal{E}_{max}/(L|\omega^2 - \omega_0^2|) = \mathcal{E}_{max}/|X_L - X_C|$

and $\delta = -90°$ for $\omega < \omega_0$ and $\delta = 90°$ for $\omega > \omega_0$.

(a) From Kirchhoff's law, $L(dI/dt) + Q/C = \mathcal{E} = \mathcal{E}_{max}\cos \omega t$. But $I = dQ/dt$, so $L\dfrac{d^2Q}{dt^2} + \dfrac{Q}{C} = \mathcal{E}_{max}\cos \omega t$.

(b) If $Q = Q_{max}\cos \omega t$ then $d^2Q/dt^2 = -\omega^2 Q$. So the result of (a) can be written $Q(1/C - L\omega^2) = \mathcal{E}$, and dividing both sides by L and recalling that $1/LC = \omega_0^2$ one obtains $Q_{max} = \mathcal{E}_{max}/[L(\omega_0^2 - \omega^2)]$.

(c) $I = dQ/dt = -\omega Q_{max}\sin \omega t = [(\omega \mathcal{E}_{max}/L)/(\omega^2 - \omega_0^2)]\sin \omega t$. Let $I_{max} = [(\omega \mathcal{E}_{max}/L)/|\omega^2 - \omega_0^2|] = \mathcal{E}_{max}/|X_L - X_C|$. Then if $\omega > \omega_0$, $I = I_{max}\sin \omega t = I_{max}\cos(\omega t - \delta)$, and if $\omega < \omega_0$, $I = -I_{max}\sin \omega t = I_{max}\cos(\omega t + \delta)$, where $\delta = -90°$.

93* •••   One method for measuring the magnetic susceptibility of a sample uses an LC circuit consisting of an air-core solenoid and a capacitor. The resonant frequency of the circuit without the sample is determined and then measured again with the sample inserted in the solenoid. Suppose the solenoid is 4.0 cm long, 0.3 cm in diameter, and has 400 turns of fine wire. Assume that the sample that is inserted in the solenoid is also 4.0 cm long and fills the air space. Neglect end effects. (In practice, a test sample of known susceptibility of the same shape as the unknown is used to calibrate the instrument.) (*a*) What is the inductance of the empty solenoid? (*b*) What should be the capacitance of the capacitor so that the resonance frequency of the circuit without a sample is 6.0000 MHz? (*c*) When a sample is inserted in the solenoid, the resonance frequency drops to 5.9989 MHz. Determine the sample's susceptibility.

(*a*) $L = \mu_0 n^2 A \ell$

$L = (4\pi \times 10^{-7})(10^4)^2(\pi \times 9 \times 10^{-6}/4)(4 \times 10^{-2})$ H $= 35.5\ \mu$H

(*b*) $4\pi^2 f_0^2 = 1/LC;\ C = (4\pi^2 f_0^2 L)^{-1}$

$C = 19.8$ pF

(*c*) $df_0/dL = -f_0/2L;\ \Delta f_0/f_0 = -\Delta L/2L;\ \Delta L = \chi L$

$\chi = -2\Delta f_0/f_0 = 3.67 \times 10^{-4}$

97* ••• At what frequency will the voltage across the load resistor of Problem 42 be half the source voltage?

1. Write $Z_p = Z$ of parallel combination of C and R_L
$Z_p = (R_p^2 + X_p^2)^{1/2}$, where $R_p = R_L X_C^2/(R_L^2 + X_C^2)$ and $X_p = -R_L^2 X_C/(R_L^2 + X_C^2)$

2. Write $Z_T = Z$ of the circuit
$Z_T = [(R + R_p)^2 + X_p^2]^{1/2} = \dfrac{\sqrt{\left(RR_L^2 + R_L X_C^2\right)^2 + R_L^4 X_C^2}}{R_L^2 + X_C^2}$

3. If $V_p = \mathcal{E}/2$ then we must have $Z_p = Z_T/2$ or $2Z_p = Z_T$
$R^2 R_L^4 + 2RR_L^3 X_C^2 + R_L^2 X_C^4 + R_L^4 X_C^2 = 4R_L^2 X_C^4 + 4R_L^4 X_C^2$

4. Substitute numerical values and solve for X_C^2
$X_C^2 = 6.048\ \Omega^2;\ X_C = 2.459\ \Omega = 1/\omega C$

5. Evaluate $f = \omega/2\pi$ with $C = 8\ \mu$F
$f = 8.09$ kHz

101*• True or false: If a transformer increases the current, it must decrease the voltage.

True

105*• The primary of a step-down transformer has 250 turns and is connected to a 120-V-rms line. The secondary is to supply 20 A at 9 V. Find (*a*) the current in the primary and (*b*) the number of turns in the secondary, assuming 100% efficiency.

(*a*) $I_1/I_2 = V_2/V_1$

$I_1 = 20(9/120) = 1.5$ A

(*b*) $N_2/N_1 = V_2/V_1$

$N_2 = 250(9/120) = 18.75 \approx 19$ turns

109*•• One use of a transformer is for *impedance matching*. For example, the output impedance of a stereo amplifier is matched to the impedance of a speaker by a transformer. In Equation 31-67, the currents I_1 and I_2 can be related to the impedance Z in the secondary since $I_2 = V_2/Z$. Using Equations 31-65 and 31-66, show that

$$I_1 = \mathcal{E}/[(N_1/N_2)^2 Z]$$

and, therefore, $Z_{\text{eff}} = (N_1/N_2)^2 Z$.

$Z = V_2/I_2$. $V_2 = (N_2/N_1)\mathcal{E}$ and $I_2 = (N_1/N_2)I_1$. So $Z = (N_2/N_1)^2 \mathcal{E}/I_1$ or $Z_{\text{eff}} = \mathcal{E}/I_1 = Z(N_1/N_2)^2$.

113*• Sketch a graph of X_L versus f for $L = 3$ mH.

The graph is shown on the right. Here X_L is in Ω and f is in Hz.

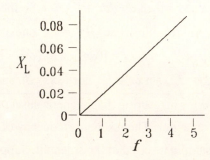

117*•• A pulsed current has a constant value of 15 A for the first 0.1 s of each second and is then 0 for the next 0.9 s of each second. (*a*) What is the rms value for this current waveform? (*b*) Each current pulse is generated by a voltage pulse of maximum value 100 V. What is the average power delivered by the pulse generator?

(*a*) $I_{rms} = (<I^2>)^{\frac{1}{2}}$, where $<>$ denotes the time average $I_{rms} = (0.1 \times 225/1.0)^{\frac{1}{2}} = 4.74$ A

(*b*) $P_{av} = I_{rms} V_{rms}$; $V_{rms} = (<V^2>)^{\frac{1}{2}}$ $V_{rms} = 31.6$ V; $P_{av} = 150$ W

121*•• Repeat Problem 120 if the resistor *R* is replaced by a 2-μF capacitor.

1. Write Kirchhoff's law equation

2. Let $q(t)/C = A \cos \omega t + B$

3. $I = dq/dt = -AC\omega \sin \omega t$

$(20 \cos \omega t + 18)$ V $= q(t)/C$

This is a steady state solutions for $A = 20$ V, $B = 18$ V

$I = -(45.2 \sin \omega t)$ mA; $I_{max} = 45.2$ mA; $I_{min} = -45.2$ mA;

$I_{rms} = 32.0$ mA

CHAPTER 32

Maxwell's Equations and Electromagnetic Waves

1* • A parallel-plate capacitor in air has circular plates of radius 2.3 cm separated by 1.1 mm. Charge is flowing onto the upper plate and off the lower plate at a rate of 5 A. (*a*) Find the time rate of change of the electric field between the plates. (*b*) Compute the displacement current between the plates and show that it equals 5 A.

(*a*) Use Equ. 23-25: $E = Q/\epsilon_0 A$; $dE/dt = (dQ/dt)/\epsilon_0 A$ $dE/dt = I/\epsilon_0 A = 3.40 \times 10^{14}$ V/m·s

(*b*) Use Equ. 32-3: $\phi_e = EA$ $I_d = \epsilon_0 A(dE/dt) = I = 5$ A

5* •• Current of 10 A flows into a capacitor having plates with areas of 0.5 m². (*a*) What is the displacement current between the plates? (*b*) What is dE/dt between the plates for this current? (*c*) What is the line integral of $B \cdot d\ell$ around a circle of radius 10 cm that lies within and parallel to the plates?

(*a*) See Problem 1 $I_d = 10$ A

(*b*) $dE/dt = I_d/\epsilon_0 A$ (see Problem 1) $dE/dt = 2.26 \times 10^{12}$ V/m·s

(*c*) Use Equ. 32-4; I_d enclosed $= I_d(\pi r^2/A)$ $\oint B \cdot d\ell = \mu_0 I_d(\pi r^2/A) = 7.90 \times 10^{-7}$ T·m

---

9* ••• In this problem, you are to show that the generalized form of Ampère's law (Equation 32-4) and the Biot–Savart law give the same result in a situation in which they both can be used. Figure 32-12 shows two charges $+Q$ and $-Q$ on the x axis at $x = -a$ and $x = +a$, with a current $I = -dQ/dt$ along the line between them. Point P is on the y axis at $y = R$. (*a*) Use the Biot–Savart law to show that the magnitude of B at point P is

$$B = \frac{\mu_0 I a}{2\pi R} \frac{1}{\sqrt{R^2 + a^2}}$$

(*b*) Consider a circular strip of radius r and width dr in the yz plane with its center at the origin. Show that the flux of the electric field through this strip is

$$E_x dA = \frac{Q}{\epsilon_0} a(r^2 + a^2)^{-3/2} r \, dr$$

(*c*) Use your result for part (*b*) to find the total flux ϕ_e through a circular area of radius R. Show that

$$\epsilon_0 \phi_e = Q\left(1 - \frac{a}{\sqrt{a^2 + R^2}}\right)$$

(d) Find the displacement current I_d, and show that

$$I + I_d = I\frac{a}{\sqrt{a^2 + R^2}}$$

(e) Then show that Equation 32-4 gives the same result for B as that found in part (a).

(a) Use Equ. 29-11 to find B at point P. Note that $\sin \theta_1 = \sin \theta_2 = \dfrac{R}{\sqrt{R^2 + a^2}}$, so $B = \dfrac{\mu_0 I}{2\pi R}\dfrac{1}{\sqrt{R^2 + a^2}}$.

(b) $E_x = [2kQ/(R^2 + a^2)]\cos \theta_1 = \dfrac{2kQa}{(r^2 + a^2)^{3/2}}$. $dA = 2\pi r\,dr$, so $E_x\,dA = \dfrac{Qa}{\epsilon_0(r^2 + a^2)^{3/2}}r\,dr$.

(c) $\epsilon_0 \phi_e = 2\pi\displaystyle\int_0^R E_x r\,dr = Qa\left(\dfrac{-1}{\sqrt{R^2 + a^2}} + \dfrac{1}{a}\right) = Q\left(1 - \dfrac{a}{\sqrt{R^2 + a^2}}\right)$.

(d) $I_d = \epsilon_0(d\phi_e/dt)$. Only Q depends on t, and $dQ/dt = -I$. So $I_d = -I\left(1 - \dfrac{a}{\sqrt{R^2 + a^2}}\right)$ and $I + I_d = \dfrac{Ia}{\sqrt{R^2 + a^2}}$.

(e) $\oint B\cdot d\ell = 2\pi RB = \mu_0(I + I_d)$; so $B = \dfrac{\mu_0 I}{2\pi R}\dfrac{1}{\sqrt{R^2 + a^2}}$; Q.E.D.

13* •   Are the frequencies of ultraviolet radiation greater or less than those of infrared radiation?

$f_{uv} > f_{ir}$

17* •   What is the frequency of an X ray with a wavelength of 0.1 nm?

$f = c/\lambda$ $f = 3\times 10^8/10^{-10} = 3\times 10^{18}$ Hz

---

21* ••   (a) For the situation described in Problem 20, at what angle is the intensity at $r = 5$ m equal to I_1? (b) At what distance is the intensity equal to I_1 at $\theta = 45°$?

(a) $1/r_1^2 = (\sin^2 \theta)/r^2$ $\sin^2 \theta = 1/4$; $\theta = 30°$

(b) $1/r_1^2 = (\sin^2 45°)/r^2$ $r^2 = 100/2 = 50$ m²; $r = 7.07$ m

---

25* •••   A small private plane approaching an airport is flying at an altitude of 2500 m above ground. The airport's flight control system transmits 100 W at 24 MHz, using a vertical dipole antenna. What is the intensity of the signal at the plane's receiving antenna when the plane's position on a map is 4 km from the airport?

From Problem 23, $I = (3P/8\pi)(\sin^2 \theta)/r^2$; $\theta = \tan^{-1}(2.5/4.0)$; $I = 0.151\ \mu W/m^2$

29* •   Show that the units of $E = cB$ are consistent; that is, show that when B is in teslas and c is in meters per second, the units of cB are volts per meter or newtons per coulomb.

$[c][B] = [m/s][N/A\cdot m] = [N/C] = [V/m]$.

33* ••   An AM radio station radiates an isotropic sinusoidal wave with an average power of 50 kW. What are the amplitudes of E_{max} and B_{max} at a distance of (a) 500 m, (b) 5 km, and (c) 50 km?

(a) $I = P_{av}/4\pi r^2 = c\epsilon_0 E_{rms}^2 = c\epsilon_0 E_{max}^2/2$; $B_{max} = E_{max}/c$ $E_{max} = 3.46$ V/m; $B_{max} = 11.5$ nT

(b) 5 km $= 10\times 500$ m; $E_{max} \propto 1/r$ $E_{max} = 0.346$ V/m; $B_{max} = 1.15$ nT

(c) 50 km = 100×500 m

E_{max} = 0.0346 V/m; B_{max} = 0.115 nT

37* ••   Instead of sending power by a 750-kV, 1000-A transmission line, one desires to beam this energy via an electromagnetic wave. The beam has a uniform intensity within a cross-sectional area of 50 m². What are the rms values of the electric and the magnetic fields?

1. Determine the intensity; $I = P/A$

$I = 7.5\times10^8/50$ W/m² = 1.5×10^7 W/m²

2. $I = E_{rms}^2/c\mu_0$; solve for E_{rms}; $B_{rms} = E_{rms}/c$

E_{rms} = 75.2 kV/m; B_{rms} = 0.251 mT

41* ••   A 10- by 15-cm card has a mass of 2 g and is perfectly reflecting. The card hangs in a vertical plane and is free to rotate about a horizontal axis through the top edge. The card is illuminated uniformly by an intense light that causes the card to make an angle of 1° with the vertical. Find the intensity of the light.

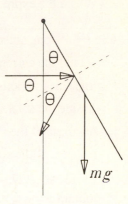

The physical arrangement is shown in the figure. Note that the force exerted by the radiation acts along the dashed line. Let the force acting on an area $dA = w\,dx$ be dF_L. The torque on an area $dA = w\,dx$ about the pivot is $d\tau_R = dF_L\,x$. Next, note that $dF_L = 2(I/c)(\cos\theta)\,dA$, the factor 2 arising from the mirror reflection. The net torque due to the radiation about the pivot is obtained by integrating $d\tau$ over the length of the card. Thus $\tau_R = (IA\ell/c)\cos\theta$. The restoring torque due to the gravitational force mg is $(mg\ell/2)\sin\theta$. Equating these torques gives $I = (mgc/2A)\tan\theta$. Substituting appropriate numerical values one finds $I = 3.42$ MW/m²

45* ••   A very long wire of radius 4 mm is heated to 1000 K. The surface of the wire is an ideal blackbody radiator. (a) What is the total power radiated per unit length? Find (b) the magnitude of the Poynting vector S, (c) E_{rms}, and (d) B_{rms} at a distance of 25 cm from the wire.

(a) Use Equ. 21-20; assume T_0 = 293 K

P_{net}/L = 1415 W/m

(b) Use Equ. 32-9; $I = P_{net}/2\pi rL = S$

S = 901 W/m²

(c) From Equs. 32-9 and 32-7, $S = E_{rms}^2/\mu_0 c$

E_{rms} = 583 V/m

(d) Use Equ. 32-7

B_{rms} = 1.94 μT

49* •   Use the known values of μ_0 and ϵ_0 in SI units to compute $c = 1/\sqrt{\epsilon_0\mu_0}$ and show that it is approximately 3×10^8 m/s.

Evaluate $(\epsilon_0\mu_0)^{-\frac{1}{2}}$

$(8.85\times10^{-12}\times4\pi\times10^{-7})^{-\frac{1}{2}} = 3.00\times10^8$ m/s

---

53* ••   A loop antenna that may be rotated about a vertical axis is used to locate an unlicensed amateur radio transmitter. If the output of the receiver is proportional to the intensity of the received signal, how does the output of the receiver vary with the orientation of the loop antenna?

The current induced in a loop antenna is proportional to the time-varying magnetic field. For maximum signal, the antenna's plane should make an angle $\theta = 0°$ with the line from the antenna to the transmitter. For any other angle, the induced current is proportional to cos θ. The intensity of the signal is therefore proportional to cos θ.

57* ••   A circular capacitor of radius a has a thin wire of resistance R connecting the centers of the two plates. A voltage $V_0\sin\omega t$ is applied between the plates. (a) What is the current drawn by this capacitor? (b) What is the magnetic field as a function of radial distance r from the centerline within the plates of this capacitor? (c) What is the phase angle between current and applied voltage?

(a) $I = I_c + I_d$; $I_c = V/R$; For I_d, use Equ. 32-4; $I_c = (V_0/R)\sin \omega t$; $I_c = (V_0 \epsilon_0 \pi \omega a^2/d)\cos \omega t$

$I_d = \epsilon_0 A(dE/dt) = (\epsilon_0 A/d)(dV/dt)$; $I_c = V/R$ $I = V_0[(1/R)\sin \omega t + (\epsilon_0 \pi \omega a^2/d)\cos \omega t]$

(b) Use Equ. 32-4; here $I_d' = I_d(r^2/a^2)$ $B(r) = (\mu_0 V_0/2 \pi r)[(1/R)\sin \omega t + (\epsilon_0 \pi r^2 \omega/d)\cos \omega t]$

(c) $\delta = \tan^{-1}(I_d/I_c)$ $\delta = \tan^{-1}(\pi a^2 \epsilon_0 \omega R/d)$

61* ••   At the surface of the earth, there is an approximate average solar flux of 0.75 kW/m². A family wishes to construct a solar energy conversion system to power their home. If the conversion system is 30% efficient and the family needs a maximum of 25 kW, what effective surface area is needed for perfectly absorbing collectors?

Write the expression for P; $P = IA\epsilon$ $A = P/I\epsilon = 111$ m²

65* •••   A long solenoid of n turns per unit length has a current that slowly increases with time. The solenoid has radius r, and the current in the windings has the form $I(t) = at$. (a) Find the induced electric field at a distance $r < R$ from the solenoid axis. (b) Find the magnitude and direction of the Poynting vector S at the cylindrical surface $r = R$ just inside the solenoid windings. (c) Calculate the flux $\oint S_n\, dA$ into the solenoid, and show that it equals the rate of increase of the magnetic energy inside the solenoid. (Here S_n is the *inward* component of S perpendicular to the surface of the solenoid.)

(a) From Equ. 29-9, $B = \mu_0 nI = \mu_0 nat$, and $\phi_m = \mu_0 nat \pi r^2$. Now apply Equ. 32-6c, i.e., $2\pi rE = -\mu_0 na \pi r^2$ and obtain $E = -\mu_0 nar/2$.

(b) At $r = R$, $S = EB/\mu_0 = \mu_0 n^2 a^2 Rt/2$. Since the field E is tangential and directed so as give an induced current that opposes the increase in B, $E \times B$ is a vector that points toward the axis of the solenoid.

(c) Consider a cylindrical surface of length L and radius R. Since S points inward, the energy flowing into the solenoid per unit time is $2\pi RLS = \pi\mu_0 n^2 a^2 tR^2 L$. The magnetic energy is $U_B = (\pi R^2 L)(B^2/2\mu_0) = (\pi R^2 L)(\mu_0 n^2 a^2 t^2/2)$ and $dU_B/dt = \pi\mu_0 n^2 a^2 tR^2 L$. Q.E.D.

69* •••   When an electromagnetic wave is reflected at normal incidence on a perfectly conducting surface, the electric field vector of the reflected wave at the reflecting surface is the negative of that of the incident wave. (a) Explain why this should be. (b) Show that the superposition of incident and reflected waves results in a standing wave. (c) What is the relationship between the magnetic field vector of the incident and reflected waves at the reflecting surface?

(a) At a perfectly conducting surface, $E = 0$. Therefore, the sum of the electric field of the incident and reflected wave must add to zero, and so $E_i = -E_r$.

(b) Let $E_i = E_{0y} \cos(\omega t - kx)$. Then $E_r = -E_{0y} \cos(\omega t + kx)$. Using $\cos(\alpha + \beta) = \cos \alpha \cos \beta - \sin \alpha \sin \beta$, $E_i + E_r = 2E_{0y} \sin(\omega t) \sin(kx)$, which is the expression for a standing wave.

(c) Using $E \times B = \mu_0 S$ and S the direction of propagation of the wave, we see that for the incident wave, $B_i = B_z \cos(\omega t - kx)$. Since both S and E_y are reversed for the reflected wave, $B_r = B_z \cos(\omega t + kx)$. So the magnetic field vectors are in the same direction at the reflecting surface and add at that surface; i.e., $B = 2B_i$.

CHAPTER 33

Properties of Light

1* • Why is helium needed in a helium–neon laser? Why not just use neon?

The population inversion between the state $E_{2,Ne}$ and the state 1.96 eV below it (see Figure 33-9) is achieved by inelastic collisions between neon atoms and helium atoms excited to the state $E_{2,He}$.

5* • The first excited state of an atom of a gas is 2.85 eV above the ground state. (a) What is the wavelength of radiation for resonance absorption? (b) If the gas is irradiated with monochromatic light of 320 nm wavelength, what is the wavelength of the Raman scattered light?

(a) Use Equ. 33-2 $\lambda = 1240/2.85$ nm = 435 nm

(b) $E_{Raman} = E_{inc} - \Delta E$; $\lambda_{Raman} = 1240/E_{Raman}$ $E_{Raman} = (1240/320 - 2.85)\,eV = 1.025\,eV; \lambda_{Raman} = 1210$ nm

---

9* • Estimate the time required for light to make the round trip in Galileo's experiment to determine the speed of light.

$\Delta t = D/c$; D = 6 km (see Problem 14) $\Delta t = 6 \times 10^3/3 \times 10^8$ s = 20 μs

---

13*• The distance from a point on the surface of the earth to one on the surface of the moon is measured by aiming a laser light beam at a reflector on the surface of the moon and measuring the time required for the light to make a round trip. The uncertainty in the measured distance Δx is related to the uncertainty in the time Δt by $\Delta x = c\,\Delta t$. If the time intervals can be measured to ± 1.0 ns, find the uncertainty of the distance in meters.

$\Delta x = \pm c\,\Delta t$ $\Delta x = \pm 3 \times 10^8 \times 10^{-9}$ m = ± 30 cm

---

17*•• The density of the atmosphere decreases with height, as does the index of refraction. Explain how one can see the sun after it has set. Why does the setting sun appear flattened?

The change in atmospheric density results in refraction of the light from the sun, bending it toward the earth. Consequently, the sun can be seen even after it is just below the horizon. Also, the light from lower portion of the sun is refracted more than that from the upper portion, so the lower part appears to be slightly higher in the sky. The effect is an apparent flattening of the disk into an ellipse.

21* • Find the speed of light in water and in glass.

Use Equ. 33-7 $v_{water} = c/1.333 = 2.25 \times 10^8$ m/s; $v_{glass} = c/1.5 = 2 \times 10^8$ m/s

25* ••   Light is incident normally on a slab of glass with an index of refraction $n = 1.5$. Reflection occurs at both
surfaces of the slab. About what percentage of the incident light energy is transmitted by the slab?
We shall neglect multiple reflections at glass-air interfaces.

1. Use Equ. 33-11 to find I in glass	$I_{glass} = I_0(1 - 0.5^2/2.5^2) = 0.96I_0$
2. Use Equ. 33-11 to find I transmitted	$I_{transm} = I_{glass}(1 - 0.5^2/2.5^2) = 0.96^2I_0 = 92.2\% \, I_0$

29* •••   Figure 33-52 shows a beam of light incident on a glass plate of thickness d and index of refraction n. (a) Find
the angle of incidence such that the perpendicular separation between the ray reflected from the top surface and that
reflected from the bottom surface and exiting the top surface is a maximum. (b) What is this angle of incidence if
the index of refraction of the glass is 1.60? What is the separation of the two beams if the thickness of the glass
plate is 4.0 cm?

(a) Let x be the perpendicular separation between the two rays. The separation between the points of emergence of
the two rays on the glass surface is $2d \tan \theta_r$, where θ_r is the angle of refraction. Then $x = 2d \tan \theta_r \cos \theta_i$, where θ_i
is the angle of incidence. To find θ_i for maximum x we differentiate x with respect to θ_i and set the derivative equal
to zero. $dx/d\theta_i = 2d[-\tan \theta_r \sin \theta_i + \sec^2 \theta_r \cos \theta_i (d\theta_r/d\theta_i)]$. From Snell's law, $\cos \theta_i \, d\theta_i = n \cos \theta_r \, d\theta_r$, or
$d\theta_r/d\theta_i = (1/n)(\cos \theta_i/\cos \theta_r)$. Thus, $dx/d\theta_i = 2d[(1/n)(\cos^2 \theta_i/\cos^3 \theta_r) - (\sin \theta_i \sin \theta_r)/\cos \theta_r]$. Using Snell's law
and $\sin^2 \theta + \cos^2 \theta = 1$, and factoring out $1/(n^3\cos^3 \theta_r)$ one obtains $dx/d\theta_i = (2d/n^3\cos^3 \theta_r)(\sin^4 \theta_i - 2n^2 \sin^2 \theta_i + n^2)$.
Setting the quantity in the second parenthesis equal to zero and solving for $\sin \theta_i$ one obtains the result

$$\sin \theta_i = n \sqrt{1 - \sqrt{1 - \frac{1}{n^2}}} \quad \text{or} \quad \theta_i = \sin^{-1}\left(n \sqrt{1 - \sqrt{1 - \frac{1}{n^2}}}\right).$$

(b) 1. Substitute $n = 1.60$ into the above expression	$\theta_i = 48.5°$
2. Find θ_r using Snell's law	$\theta_r = \sin^{-1}[(\sin 48.5°)/1.6] = 27.9°$
3. $x = 2d \tan \theta_r \cos \theta_i$	$x = 2.81$ cm

33* •   What is the critical angle for total internal reflection for light traveling initially in water that is incident on a
water–air interface?

Use Equ. 33-12	$\theta_c = \sin^{-1}(1/1.333) = 48.6°$

37* ••   A point source of light is located at the bottom of a steel tank, and an opaque circular card of radius 6.0 cm is
placed over it. A transparent fluid is gently added to the tank such that the card floats on the surface with its center
directly above the light source. No light is seen by an observer above the surface until the fluid is 5 cm deep. What
is the index of refraction of the fluid?

1. Determine θ_c from the geometry	$\theta_c = \tan^{-1}(6/5) = 50.2°$
2. Use Equ. 33-12 to find n	$n = 1/\sin 50.2° = 1.30$

41* •••   A laser beam is incident on a plate of glass of thickness 3 cm. The glass has an index of refraction of 1.5 and
the angle of incidence is $40°$. The top and bottom surfaces of the glass are parallel and both produce reflected
beams of nearly the same intensity. What is the perpendicular distance d between the two adjacent reflected beams?

1. Use Snell's law to find θ_r	$\theta_r = \sin^{-1}[(\sin 40°)/1.5] = 25.37°$
2. $x = 2d \tan \theta_r \cos \theta_i$ (see Problem 29); find θ_r	$x = 2.18$ cm

45*• Two polarizers have their transmission axes at an angle θ. Unpolarized light of intensity I is incident upon the first polarizer. What is the intensity of the light transmitted by the second polarizer? (a) $I \cos^2 \theta$ (b) $(I \cos^2 \theta)/2$ (c) $(I \cos^2 \theta)/4$ (d) $I \cos \theta$ (e) $(I \cos \theta)/4$ (f) None of the above.

(b)

49*• Two polarizing sheets have their transmission axes crossed so that no light gets through. A third sheet is inserted between the first two such that its transmission axis makes an angle θ with that of the first sheet. Unpolarized light of intensity I_0 is incident on the first sheet. Find the intensity of the light transmitted through all three sheets if (a) $\theta = 45°$ and (b) $\theta = 30°$.

Let I_n be the intensity after the n'th polarizing sheet.

(a) Find I_1, I_2, and I_3; use Equ. 33-20 for I_2 and I_3 $I_1 = I_0/2$; $I_2 = I_1/2 = I_0/4$; $I_3 = I_2/2 = I_0/8$

(b) Repeat with $\theta_{1,2} = 30°$, $\theta_{2,3} = 60°$ $I_1 = I_0/2$; $I_2 = 3I_1/4 = 3I_0/8$; $I_3 = I_2/4 = 3I_0/32$

53*•• A stack of $N + 1$ ideal polarizing sheets is arranged with each sheet rotated by an angle of $\pi/2N$ rad with respect to the preceding sheet. A plane linearly polarized light wave of intensity I_0 is incident normally on the stack. The incident light is polarized along the transmission axis of the first sheet and therefore perpendicular to the transmission axis of the last sheet in the stack. (a) What is the transmitted intensity through the stack? (b) For 3 sheets ($N = 2$), what is the transmitted intensity? (c) For 101 sheets, what is the transmitted intensity? (d) What is the direction of polarization of the transmitted beam in each case?

(a) 1. Find the ratio I_{n+1}/I_n $I_{n+1}/I_n = \cos^2(\pi/2N)$

     2. There are N such reductions of intrensity $I_{N+1}/I_1 = I_{N+1}/I_0 = \cos^{2N}(\pi/2N)$; $I_{N+1} = I_0 \cos^{2N}(\pi/2N)$

(b) Find I_3 from part (a); see also Problem 49(a) $I_3 = I_0 \cos^4(\pi/4) = I_0/4$

(c) Repeat part (b) for $N = 100$ $I_{101} = I_0 \cos^{200}(\pi/200) = 0.976$

(d) In each case, the polarization of the transmitted beam is perpendicular to that of the incident beam.

57*•• Show that the electric field of a circularly polarized wave propagating in the x direction can be expressed by

$$E = E_0 \sin(kx - \omega t)\, j + E_0 \cos(kx - \omega t)\, k$$

Note that $E_y = E_0 \sin \theta$ and $E_x = E_0 \cos \theta$, where $\theta = kx - \omega t$. Clearly, The magnitude of E is constant in time and in a plane perpendicular to the direction of propagation the E vector rotates with angular frequency ω.

61*•• Vertically polarized light of intensity I_0 is incident on a stack of N ideal polarizing sheets whose angles with respect to the vertical are $\theta_n = n\pi/2N$. Determine the direction of polarization of the transmitted light and its intensity. Show that as $N \to \infty$ the direction of polarization is rotated without loss of intensity.

From Problem 53(a), $I_N = I_0 \cos^{2N}(\pi/2N)$, and as in Problem 53(d) the direction of polarization has been rotated by 90°. For $N \to \infty$ use the small angle expansion $\cos \theta = 1 - \theta^2/2 + \cdots$. Thus, $I_N = I_0(1 - \pi^2/8N^2)^{2N} = I_0(1 - \pi^2/4N)$, and as $N \to \infty$, $I_N = I_0$.

65*• A beam of monochromatic red light with a wavelength of 700 nm in air travels in water. (a) What is the wavelength in water? (b) Does a swimmer underwater observe the same color or a different color for this light?

(a) $\lambda_n = v/f = c/nf$; $\lambda_n = \lambda_0/n$ $\lambda_{water} = 700/1.333$ nm $= 525$ nm

(b) Color observed depends on frequency   Swimmer observed same color in water and air

69*••   A silver coin sits on the bottom of a swimming pool that is 4 m deep. A beam of light reflected from the coin emerges from the pool making an angle of 20° with respect to the water's surface and enters the eye of an observer. Draw a ray from the coin to the eye of the observer. Extend this ray, which goes from the water–air interface to the eye, straight back until it intersects with the vertical line drawn through the coin. What is the apparent depth of the swimming pool to this observer?

The sketch shows the ray from the coin through the water and to the eye of the observer. The angles shown are as determined below.

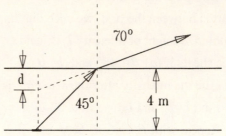

1. Find the angle θ_i at the water-air interface $\theta_i = \sin^{-1}[(\sin 70°)/1.33] = 45°$

2. Find d $d = (4 \times \tan 45° \times \tan 20°)\ m = 1.46\ m$

73* ••   Show that when a mirror is rotated through an angle θ, the reflected beam of light is rotated through 2θ.

Let α be the initial angle of incidence. Since the angle of reflection with the normal to the mirror is also α, the angle between incident and reflected rays is 2α. If the mirror is now rotated by a further angle θ, the angle of incidence is increased by θ to $\alpha + \theta$, and so is the angle of reflection. Consequently, the reflected beam is rotated by 2θ relative to the incident beam.

77* ••   Light passes symmetrically through a prism having an apex angle of α as shown in Figure 33-58. (*a*) Show that the angle of deviation δ is given by

$$\sin \frac{\alpha + \delta}{2} = n \sin \frac{\alpha}{2}$$

(*b*) If the refractive index for red light is 1.48 and that for violet light is 1.52, what is the angular separation of visible light for a prism with an apex angle of 60°?

(*a*) With respect to the normal to the left face of the prism, let the angle of incidence be θ_i and the angle of refraction be θ_r. From the geometry of Figure 33-58 it is evident that $\theta_r = \alpha/2$. The angle of deviation at this refracting interface is $\theta_i - \alpha/2$. By symmetry, the angle of deviation at the second refracting interface is also of this magnitude. Thus, $\delta = 2\theta_i - \alpha$ and $\theta_i = \frac{1}{2}(\alpha + \delta)$. Using Snell's law, $\sin \theta_i = n \sin (\alpha/2)$, we obtain the stated expression, $\sin \frac{1}{2}(\alpha + \delta) = n \sin \frac{1}{2}\alpha$.

(*b*) 1. $\delta = 2 \sin^{-1}[n \sin \frac{1}{2}\alpha] - \alpha$. Find δ_{violet} and δ_{red} $\delta_{violet} = 38.93°;\ \delta_{red} = 35.46°$

   2. Angular separation $= \delta_{violet} - \delta_{red}$ Angular separation $= 3.47°$

81*••   Given that the index of refraction for red light in water is 1.3318 and that the index of refraction for blue light in water is 1.3435, find the angular separation of these colors in the primary rainbow. (Use the equation given in Problem 86.)

1. Find θ_{1m} for red and blue light; $\cos \theta_{1m} = \sqrt{\dfrac{n^2 - 1}{3}}$ Red light: $\theta_{1m} = 59.48°$; blue light: $\theta_{1m} = 58.80°$

2. Use Equ. 33-18 to find ϕ_d for red and blue light

Red light: $\phi_d = 137.75°$; blue light: $\phi_d = 139.42°$

3. Angular separation = $\phi_{d,blue} - \phi_{d,red}$

Angular separation = 1.67°

85* •• Suppose rain falls vertically from a stationary cloud 10,000 m above a confused marathoner running in a circle with constant speed of 4 m/s. The rain has a terminal speed of 9 m/s. (*a*) What is the angle that the rain appears to make with the vertical to the marathoner? (*b*) What is the apparent motion of the cloud as observed by the marathoner? (*c*) A star on the axis of the earth's orbit appears to have a circular orbit of angular diameter of 41.2 seconds of arc. How is this angle related to the earth's speed in its orbit and the velocity of photons received from this distant star? (*d*) What is the speed of light as determined from the data in part (*c*)?

(*a*) $\tan\theta = v_{run}/v_{rain}$; $\theta = \tan^{-1}(v_{run}/v_{rain})$ 　　　　　$\theta = \tan^{-1}(4/9) = 24.0°$

(*b*) Cloud moves in circle of radius $R = H\tan\theta$ 　　　　Radius of circle of cloud motion is $R = 4444$ m

(*c*) As in (*a*) where $v_{run} = v_{earth}$, $v_{rain} = c$ 　　　$\tan\theta = v_{earth}/c$; $\theta = ½×$(angular diameter)

(*d*) 1. Find $v_{earth} = 2\pi R_{earth-sun}/(1\ y)$ 　　$v_{earth} = (2\pi×1.5×10^{11}/3.156×10^7)$ m/s $= 2.986×10^4$ m/s

　　　2. $c = v_{earth}/\tan\theta$ 　　　　　　$c = [2.986×10^4/\tan(20.6'')]$ m/s $= 2.99×10^8$ m/s

---

89*••• Show that the angle of deviation δ is a minimum if the angle of incidence is such that the ray passes through the prism symmetrically as shown in Figure 33-58.

The figure to the right shows the prism and the path of the ray through it. The light dashed lines are the normals to the prism faces. The triangle formed by the interior ray and the prism faces has interior angles of α, $90° - \theta_2$, and $90° - \theta_3$. Consequently, $\theta_2 + \theta_3 = \alpha$. Next we note that the deviation angle $\delta = \theta_1 + \theta_4 - \alpha$. These are purely geometric relations. We now employ Snell's law to relate θ_3 to θ_4 and θ_1 to θ_2. We have $\sin\theta_1 = n\sin\theta_2$ and $\sin\theta_4 = n\sin\theta_3$. We can now write an expression for δ in terms of one of the angles θ_i and the apex angle α.

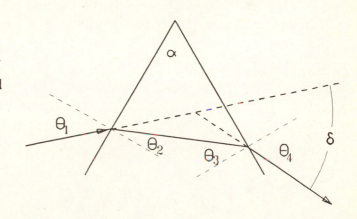

$\delta = \sin^{-1}(n\sin\theta_2) + \sin^{-1}(n\sin\theta_3) - \alpha = \sin^{-1}[n\sin(\theta_3 - \alpha)] + \sin^{-1}(n\sin\theta_3) - \alpha$. The only variable in this expression is θ_3. To determine the condition for minimizing δ we now take the derivative with respect to θ_3 and set it equal to zero.

$$\frac{d\delta}{d\theta_3} = -\frac{n\cos(\alpha - \theta_3)}{\sqrt{1 - [n\sin(\alpha - \theta_3)]^2}} + \frac{n\cos\theta_3}{\sqrt{1 - (n\sin\theta_3)^2}} = 0.$$ This equation is satisfied if $\alpha - \theta_3 = \theta_3$ or $\theta_3 = \alpha/2$.

But $\theta_2 = \alpha - \theta_3 = \alpha/2$, which shows that the deviation angle is a minimum if the ray passes through the prism symmetrically.

CHAPTER **34**

Optical Images

1* • Can a virtual image be photographed?

Yes. Note that a virtual image is "seen" because the eye focuses the diverging rays to form a real image on the retina. Similarly, the camera lens can focus the diverging rays onto the film.

5* • Two plane mirrors make an angle of 90°. Show by considering various object positions that there are three images for any position of an object. Draw appropriate bundles of rays from the object to the eye for viewing each image.

Three virtual images are formed, as shown in the adjoining figure. The eye should be to the right and above the mirrors.

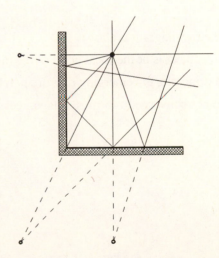

9* •• Under what condition will a concave mirror produce an erect image? A virtual image? An image smaller than the object? An image larger than the object?

If $s < f$, the image is virtual, erect, and larger than the object.

If $f < s < 2f$, the image is real, inverted, and larger than the object.

If $s > 2f$, the image is real, inverted, and smaller than the object.

13* • A concave spherical mirror has a radius of curvature of 40 cm. Draw ray diagrams to locate the image (if one is formed) for an object at a distance of (*a*) 100 cm, (*b*) 40 cm, (*c*) 20 cm, and (*d*) 10 cm from the mirror. For each case, state whether the image is real or virtual; erect or inverted; and enlarged, reduced, or the same size as the object.

(*a*) The ray diagram is shown.

The image is real, inverted, and reduced.

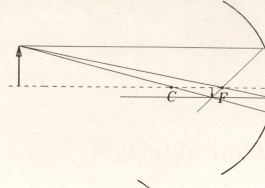

(*b*) The ray diagram is shown.

The image is real, inverted, and of the same size as the object.

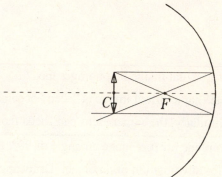

(*c*) The ray diagram is shown.

The object is at the focal point of the mirror.
In principle, the emerging rays are parallel and do not converge to form an image, real or virtual.

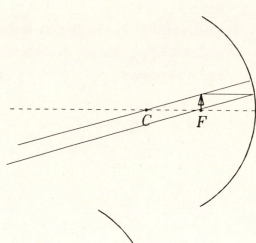

(*d*) The ray diagram is shown.

The image is virtual, upright, and enlarged.

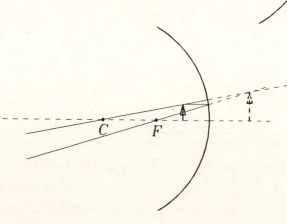

17*•   Show that a convex mirror cannot form a real image of a real object, no matter where the object is placed, by showing that s' is always negative for a positive s.

From Equ. 34-3, $1/s' = 1/f - 1/s = (s - f)/sf$. For a convex mirror, $f < 0$. With $s > 0$, the numerator is positive and the denominator negative. Consequently, $1/s'$ is negative and so is s'.

21*••   A concave spherical mirror has a radius of curvature of 6.0 cm. A point object is on the axis 9 cm from the mirror. Construct a precise ray diagram showing rays from the object that make angles of 5°, 10°, 30°, and 60° with the axis, strike the mirror, and are reflected back across the axis. (Use a compass to draw the mirror, and use a protractor to measure the angles needed to find the reflected rays.) What is the spread δx of the points where these rays cross the axis?

The rays from the point object are shown.

Note that the rays that reflect from the mirror far from the axis do not converge at the same point as those that reflect from the mirror close to the mirror axis. For the small-angle rays, the point of convergence is 4.5 cm from the mirror. The 60° ray crosses the axis at 3 cm from the mirror. Consequently, the image extends from 4.5 cm to 3.0 cm, or about 1.5 cm along the axis.

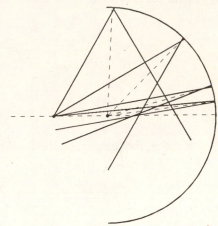

25*••   Parallel light from a distant object strikes the large mirror in Figure 34-54 ($r = 5$ m) and is reflected by the small mirror that is 2 m from the large mirror. The small mirror is actually spherical, not planar as shown. The light is focused at the vertex of the large mirror. (a) What is the radius of curvature of the small mirror? (b) Is it convex or concave?

(a) 1. Locate the image produced by the large mirror $1/s' = (1/2.5)$; $s' = 2.5$ m

   2. This serves as a virtual image for the small $1/f_{small} = (1/2 - 1/0.5)$ m^{-1}; $f_{small} = -0.667$ m

   mirror at $s = -0.5$ m. Find f_{small} and r_{small}

(b) $f < 0$ $r_{small} = 2f_{small} = -1.33$ m

   The small mirror is convex.

29* •   A fish is 10 cm from the front surface of a fish bowl of radius 20 cm. (a) Where does the fish appear to be to someone in air viewing it from in front of the bowl? (b) Where does the fish appear to be when it is 30 cm from the front surface of the bowl?

(a) 1. The object is inside the bowl; determine r $r = -20$ cm

   2. Use Equ. 34-5; $n_1 = 1.33$, $n_2 = 1.0$, $s = 10$ cm $s' = -8.58$ cm; 8.58 cm from front surface of the bowl

(b) Repeat (a) for $s = 30$ cm $s' = -35.9$ cm; 35.9 cm from front surface of the bowl

33* ••   Repeat Problem 30 when the glass rod and objects are immersed in water.

(*a*) 1. Use Equ. 34-5 with $n_1 = 1.33$, $n_2 = 1.5$ to find $s' = f = [1.5 \times 5/(1.5 - 1.33)]$ cm $= 44.1$ cm
f; set $s = \infty$, then $s' = f$

   2. Use Equ. 34-5 to find s' for $s = 20$ cm $s' = -46.2$ cm; image virtual, upright, enlarged; see below

(*b*) Repeat part (*a*) 2. for $s = 5$ cm $s' = -6.47$ cm; image virtual, upright; see below

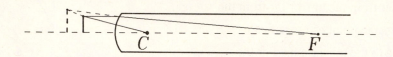

(*c*) For $s = \infty$, the image is at the focal point, i.e., $s' = 44.1$ cm, and is of zero size. See below.

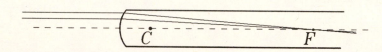

37* •   Under what conditions will the focal length of a thin lens be positive? Negative?

The lens will be positive if its index of refraction is greater than that of the surrounding medium and the lens is thicker in the middle than at the edges. Conversely, if the index of refraction of the lens is less than that of the surounding medium, the lens will be positive if it is thinner at its center than at the edges.

The lens will be negative if its index of refraction is greater than that of the surrounding medium and the lens is thinner at the center than at the edges. Conversely, if the index of refraction of the lens is less than that of the surrounding medium, the lens will be negative if it is thicker at the center than at the edges.

41* •   The following thin lenses are made of glass with an index of refraction of 1.5. Make a sketch of each lens, and find its focal length in air: (*a*) double convex, $r_1 = 10$ cm and $r_2 = -21$ cm; (*b*) plano-convex, $r_1 = \infty$ and $r_2 = -10$ cm; (*c*) double concave, $r_1 = -10$ cm and $r_2 = +10$ cm; (*d*) plano-concave, $r_1 = \infty$ and $r_2 = +20$ cm.

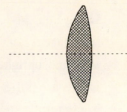

(a) Use Equ. 34-11 to find f $f = 13.5$ cm

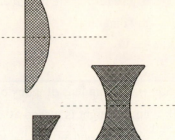

(b), (c), (d) Repeat as in (a)   (b) $f = 20$ cm

(c) $f = -10$ cm

(d) $f = -40$ cm

45* •   The following thin lenses are made of glass of index of refraction 1.6. Make a sketch of each lens, and find its focal length in air: (a) $r_1 = 20$ cm, $r_2 = 10$ cm; (b) $r_1 = 10$ cm, $r_2 = 20$ cm; (c) $r_1 = -10$ cm, $r_2 = -20$ cm.

(a) Use Equ. 34-11 to find f $f = -33.3$ cm

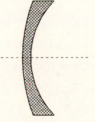

(b), (c) Repeat as in (a)   (b) $f = 33.3$ cm

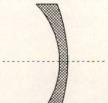

(c) $f = -33.3$ cm

49* •   Repeat Problem 47 for an object 1.0 cm high placed 10 cm in front of a thin lens whose power is -20 D. The ray diagram is shown on the next page.

Using Equ. 34-12 with $f = 1/D = -5$ cm
the image distance is -3.33 cm. The image
is virtual, upright, and diminished. The
image size is $1(3.33/10)$ cm $= 0.33$ cm.
These results are in agreement with the
ray diagram.

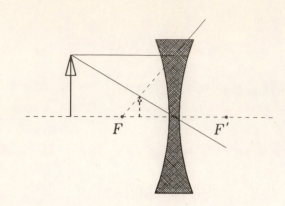

53* •• A thin lens of index of refraction 1.5 has one convex side with a radius of magnitude 20 cm. When an object 1 cm in height is placed 50 cm from this lens, an upright image 2.15 cm in height is formed. (*a*) Calculate the radius of the second side of the lens. Is it concave or convex? (*b*) Draw a sketch of the lens.

(*a*) 1. Use Equs. 34-14 and 34-12 to find f
$$f = \frac{ss'}{s + s'}; \quad s' = -2.15s; \quad f = \frac{-2.15s}{-1.15} = 93.5 \text{ cm}$$

2. Use Equ. 34-11 to find r_2; $r_2 = \dfrac{r_1 f(n-1)}{f(n-1) - r_1}$ $r_2 = 35$ cm; $r_2 > 0$, side is concave

(*b*) The lens is shown in the adjacent figure.

57* ••• In a convenient form of the thin-lens equation used by Newton, the object and image distances are measured from the focal points. Show that if $x = s - f$ and $x' = s' - f$, the thin-lens equation can be written as $xx' = f^2$, and the lateral magnification is given by $m = -x'/f = -f/x$. Indicate x and x' on a sketch of a lens.

From the definitions of x and x' it follows that Equ. 34-12 takes the form $\dfrac{1}{x + f} + \dfrac{1}{x' + f} = \dfrac{1}{f}$. Expanding this equation one obtains $f(x' + x + 2f) = xx' + xf + x'f + f^2$; $f^2 = xx'$. Q.E.D.

The lateral magnification is $m = -s'/s = -(x' + f)/(x + f)$. With $x = f^2/x'$, $m = -\dfrac{x' + f}{f(f/x' + 1)} = -\dfrac{x'}{f} = -\dfrac{f}{x}$.

The variables $x, x', f, s,$ and s' are shown in the adjacent drawing.

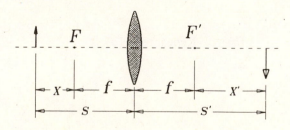

61* • True or false:

(*a*) Aberrations occur only for real images.

(*b*) Chromatic aberration does not occur with mirrors.

(*a*) False   (*b*) True

65* •• A nearsighted person who wears corrective lenses would like to examine an object at close distance. Identify the correct statement.

(*a*) The corrective lenses give an enlarged image and should be worn while examining the object.

(*b*) The corrective lenses give a reduced image of the object and should be removed.

(*c*) The corrective lenses result in a magnification of unity; it does not matter whether they are worn or removed.

(*b*)

69* •   A farsighted person requires lenses with a power of 1.75 D to read comfortably from a book that is 25 cm from the eye. What is that person's near point without the lenses?

See Example 34-13; solve for $|s'|$ = near point $f = 1/D = 57.1$ cm; $s' = -44.4$ cm; near point at 44.4 cm

---

73* ••   Since the index of refraction of the lens of the eye is not very different from that of the surrounding material, most of the refraction takes place at the cornea, where n changes abruptly from 1.0 in air to about 1.4. Assuming the cornea to be a homogeneous sphere with an index of refraction of 1.4, calculate its radius if it focuses parallel light on the retina a distance 2.5 cm away. Do you expect your result to be larger or smaller than the actual radius of the cornea?

1. Use Equ. 34-5 with $s = \infty$, $s' = 2.5$ cm to find r $r = 2.5(0.4/1.4)$ cm $= 0.714$ cm

2. The eye is not a homogeneous sphere. It is filled with a transparent liquid (vitreous humor) which has an index of refraction that is not known. If that index of refraction differs from 1.4, there is refraction at the inner surface of the cornea which will result in the formation of the image nearer the cornea's surface if $n > 1.4$ and farther if $n < 1.4$, where n is the index of refraction of the vitreous humor. If $n < 1.4$, then r as calculated above is too small.

77* •   A person with a near point distance of 30 cm uses a simple magnifier of power 20 D. What is the magnification obtained if the final image is at infinity?

Use Equ. 34-20 $M = x_{np}/f = x_{np}D = 0.3 \times 20 = 6$

81* ••   A botanist examines a leaf using a convex lens of power 12 D as a simple magnifier. What is the expected angular magnification if (*a*) the final image is at infinity, and (*b*) the final image is at 25 cm?

(*a*) Use Equ. 34-20 $M = -x_{np}D = -0.25 \times 12 = -3$

(*b*) Use Equ. 34-12 to find s; $m = -s'/s$ $1/s = 12 + 4$; $s = 1/16$; $m = -16/4 = -4$

85* ••   A microscope has an objective of focal length 16 mm and an eyepiece that gives an angular magnification of 5 for a person whose near point is 25 cm. The tube length is 18 cm. (*a*) What is the lateral magnification of the objective? (*b*) What is the magnifying power of the microscope?

(*a*) Use Equ. 34-21 $m_o = -18/1.6 = -11.25$

(*b*) $M = 5m_o$ $M = -56.25$

89* •••   A microscope has a magnifying power of 600, and an eyepiece of angular magnification of 15. The objective lens is 22 cm from the eyepiece. Without making any approximations, calculate (*a*) the focal length of the eyepiece, (*b*) the location of the object such that it is in focus for a normal relaxed eye, and (*c*) the focal length of the objective lens.

(*a*) Use Equ. 34-20 for the eyepiece; $M_e = 15$ $f_e = 25/15 = 1.67$ cm

(*b*) Use Equ. 34-22 to find $m_o = -s'/s$ for objective $m_o = -600/15 = -40$; note that $s' = (22 - 1.67)$ cm

   Find $s = -s'/m_o$ $s = 20.33/40$ cm $= 0.508$ cm

(*c*) Use Equ. 34-12 to find f_o $f_o = (20.33 \times 0.508/20.84)$ cm $= 0.496$ cm

---

93* ••   An astronomical telescope has a magnifying power of 7. The two lenses are 32 cm apart. Find the focal length of each lens.

   1. Write expressions for $M = -f_o/f_e$ and $L = f_o + f_e$ $f_o/f_e = 7; f_o + f_e = 32$ cm

   2. Solve for f_o and f_e $f_o = 28$ cm; $f_e = 4$ cm

97* •••   If you look into the wrong end of a telescope, that is, into the objective, you will see distant objects reduced in size. For a refracting telescope with an objective of focal length 2.25 m and an eyepiece of focal length 1.5 cm, by what factor is the angular size of the object reduced?

   In this case, $M = -f_e/f_o$ as the role of objective and $M = -1.5/225 = -6.67 \times 10^{-3} = 1/150$

   eyepiece are reversed

101* •   Show that a diverging lens can never form a real image from a real object. (*Hint:* Show that s' is always negative.)

   From Equ. 34-12 it follows that $s' = sf/(s - f)$. For the diverging lens, $f < 0$ and $s > 0$ for a real object. Consequently, the denominator is positive and the numerator negative, so s' must always be negative.

105* •   A thin converging lens of focal length 10 cm is used to obtain an image that is twice as large as a small object. Find the object and image distances if (*a*) the image is to be erect and (*b*) the image is to be inverted. Draw a ray diagram for each case.

   (*a*) 1. Erect image → virtual image; $m = -s'/s$ $m = 2; s' = -2s$

       2. Use Equ. 34-12 to find s and s' $s = f/2 = 5$ cm; $s' = -10$ cm

   (*b*) 2. Inverted image → real image $s' = 2s; s = 3f/2 = 15$ cm; $s' = 30$ cm

   The ray diagrams for cases (*a*) and (*b*) are shown below.

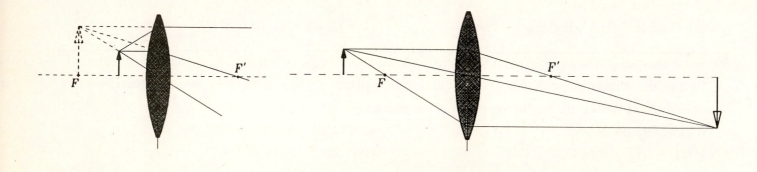

109* ••   A 35-mm camera has a picture size of 24 mm by 36 mm. It is used to take a picture of a person 175 cm tall so that the image just fills the height (24 mm) of the film. How far should the person stand from the camera if the focal length of the lens is 50 mm?

   1. Find $m = -s'/s$ $m = -2.4/175 = -1.37 \times 10^{-2}$

   2. Use Equ. 34-12 to find s $s = 0.0137s \times 5/(0.0137s - 5); s = 370$ cm $= 3.7$ m

113*•• (*a*) Find the focal length of a *thick* double-convex lens with an index of refraction of 1.5, a thickness of 4 cm, and radii of +20 cm and −20 cm. (*b*) Find the focal length of this lens in water.

(*a*) Here we must consider refraction at each surface separately, using Equ. 34-5. To find the focal length we imagine the object at $s = \infty$, find the image from the first refracting surface at s_1'. That image serves as the object for the second refracting surface. We shall find that this is a virtual image for the second refracting surface, i.e., s_2 is negative. Using Equ. 34-5 once more, we shall locate the image formed by the second refracting surface by the virtual object at s_2. The location of that image is then the focal point of the thick lens.

    1. Use Equ. 34-5 to find s_1' $s_1' = nr_1/(n-1) = 60$ cm

    2. Find $s_2 = -(s_1' - 4\text{ cm})$ $s_2 = -56$ cm

    3. Use Equ. 34-5 to find s_2' $1/s_2' = (1.5/56 + 0.5/20)\text{ cm}^{-1}$; $s_2' = 19.3$ cm

    4. f is measured from lens's center; $f = s_2' + 2$ cm $f = 21.3$ cm

(*b*) We proceed as in part (*a*) except that now $n_1 = 1.33$ for the first refraction and $n_2 = 1.33$ for the second refraction to determine the focal length in water, which we denote by f_w.

    1. Use Equ. 34-5 to find s_1' and s_2 $s_1' = 1.5 \times 20/(1.5 - 1.33)\text{ cm} = 176$ cm; $s_2 = -172$ cm

    2. Use Equ. 34-5 to find s_2' $s_2' = 77.3$ cm

    3. $f_w = s_2' + 2$ cm $f_w = 79.3$ cm

Note that if we use the expression given in Problem 112 we obtain $f_w = 83.3$ cm, in only moderate agreement with the exact result given above.

117*••• An object is 15 cm to the left of a thin convex lens of focal length 10 cm. A concave mirror of radius 10 cm is 25 cm to the right of the lens. (*a*) Find the position of the final image formed by the mirror and lens. (*b*) Is the image real or virtual? Erect or inverted? (*c*) Show on a diagram where your eye must be to see this image.

We begin with parts (*b*) and (*c*), the ray diagram for this situation. This is shown below. **1** represents the object. Two rays from **1** are shown; one passes through the center of the lens, the other is paraxial and then passes through the focal point F'. The two rays intersect behind the mirror, and the image formed there, identified as **2**, serves as a virtual object for the mirror. Two rays are shown emanating from this virtual image, one through the center of the mirror, the other passing through its focal point (half way between C and the mirror surface) and then continuing as a paraxial ray. These two rays intersect in front of the mirror, forming a real image, identified as **3**. Finally, the image **3** serves as a real object for the lens; again we show two rays, a paraxial ray that then passes through the focal point F and a ray through the center of the lens. These two rays intersect to form the final real, upright, diminished image, identified as **4**. To see this image the eye must be to the left of the image **4**.

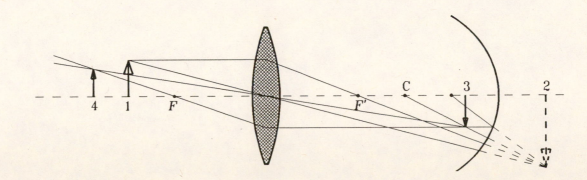

(a) 1. Use Equ. 34-12 to find the location of **2** $s' = 30$ cm; thus **2** is 5 cm behind the mirror

 2. Use Equ. 34-3 to locate **3**, where $s = -5$ cm $s' = 2.5$ cm; the image is 22.5 cm from the lens

 3. Use Equ. 34-12 to locate **4**, where $s = 22.5$ cm $s' = 18$ cm; **4** is 18 cm to the left of the lens

---

121*••• A lens with one concave side with a radius of magnitude 17 cm and one convex side with a radius of magnitude 8 cm has a focal length in air of 27.5 cm. When placed in a liquid with an unknown index of refraction, the focal length increases to 109 cm. What is the index of refraction of the liquid?

From Equ. 34-11, $n = \dfrac{r_1 r_2}{f(r_2 - r_1)} + 1$; find n $n = 1.55$

From the result of Problem 112, $n_L = \dfrac{f_L n}{f_L + f(n-1)}$ $n_L = 1.36$

125*••• A thin double-convex lens has radii r_1 and r_2 and an index of refraction n_L. The surface of radius r_1 is in contact with a liquid of index of refraction n_1, and the surface of radius r_2 is in contact with a liquid of index of refraction n_2. Show that the thin-lens equation for this situation can be expressed as $n_1/s + n_2/s' = n_2/f$ where the focal length is given by

$$\frac{1}{f} = \frac{n_L - n_1}{n_2 r_1} - \frac{n_L - n_2}{n_2 r_2}$$

Consider refraction at each surface separately. The image formed by the first refraction serves as the object for the second. We denote the object distance s_1, the image distance formed by the first surface s_1'. Then the object distance for the second surface (using the thin lens approximation) is $s_2 = -s_1'$ and the final image distance is denoted by s_2'. From Equ. 34-5 we have

$\dfrac{n_1}{s_1} + \dfrac{n_L}{s_1'} = \dfrac{n_L - n_1}{r_1}$ and $\dfrac{n_2}{s_2'} + \dfrac{n_L}{s_2} = \dfrac{n_2 - n_L}{r_2}$. But $\dfrac{n_L}{s_2} = -\dfrac{n_L}{s_1'} = \dfrac{n_1}{s_1} - \dfrac{n_L - n_1}{r_1}$. We now substitute this into the

second equation and obtain $\dfrac{n_1}{s} + \dfrac{n_2}{s'} = \dfrac{n_L - n_1}{r_1} - \dfrac{n_L - n_2}{r_2}$, where we have made the substitution $s_1 = s$ and $s_2' = s'$.

To determine the focal length, we set $s = \infty$ and solve for $s' = f$. This gives $\dfrac{1}{f} = \dfrac{n_L - n_1}{n_2 r_1} - \dfrac{n_L - n_2}{n_2 r_2}$ and, from the

previous equation we can now write the lens equation in terms of the above focal length, namely, $\dfrac{n_1}{s} + \dfrac{n_2}{s'} = \dfrac{n_2}{f}$.

CHAPTER **35**

Interference and Diffraction

1* • When destructive interference occurs, what happens to the energy in the light waves?

The energy is distributed nonuniformly in space; in some regions the energy is below average (destructive interference), in others it is higher than average (constructive interference).

5* •• Two coherent microwave sources that produce waves of wavelength 1.5 cm are in the xy plane, one on the y axis at $y = 15$ cm and the other at $x = 3$ cm, $y = 14$ cm. If the sources are in phase, find the difference in phase between the two waves from these sources at the origin.

1. Find $\Delta r = r_2 - r_1$ $r_1 = 15$ cm, $r_2 = 14.318$ cm; $\Delta r = 0.682$ cm

2. Use Equ. 35-1 $\delta = (0.682/1.5)360° = 164°$

9* • A loop of wire is dipped in soapy water and held so that the soap film is vertical. (*a*) Viewed by reflection with white light, the top of the film appears black. Explain why. (*b*) Below the black region are colored bands. Is the first band red or violet? (*c*) Describe the appearance of the film when it is viewed by *transmitted* light.

(*a*) The phase change on reflection from the front surface of the film is 180°; the phase change on reflection from the back surface of the film is 0°. As the film thins toward the top, the phase change associated with the film's thickness becomes negligible and the two reflected waves interfere destructively.

(*b*) The first constructive interference will arise when $t = \lambda/4$. Therefore, the first band will be violet (shortest visible wavelength).

(*c*) When viewed in transmitted light, the top of the film is white, since no light is reflected. The colors of the bands are those complimentary to the colors seen in reflected light; i.e., the top band will be red.

13* •• A thin film having an index of refraction of 1.5 is surrounded by air. It is illuminated normally by white light and is viewed by reflection. Analysis of the resulting reflected light shows that the wavelengths 360, 450, and 602 nm are the only missing wavelengths in or near the visible portion of the spectrum. That is, for these wavelengths, there is destructive interference. (*a*) What is the thickness of the film? (*b*) What visible wavelengths are brightest in the reflected interference pattern? (*c*) If this film were resting on glass with an index of refraction of 1.6, what wavelengths in the visible spectrum would be missing from the reflected light?

(*a*) 1. Destructive interference condition: $\lambda_m = 2nt/m$ $450/360 = (m + 1)/m$; $m = 4$ for $\lambda = 450$ nm

2. $t = m\lambda_m/2n$　　　　　　　　　　　　　　　　$t = (900/1.5)$ nm $= 600$ nm

(b) Constructive interference: $2nt/\lambda_m = m + \frac{1}{2}$　　　For $m = 2, 3$, and 4, $\lambda_m = 720$ nm, 514 nm, and 400 nm,

　　　　　　　　　　　　　　　　　　　　　　　respectively. These are the only λ's in the visible range.

(c) Now destructive interference for $2nt/\lambda_m = m + \frac{1}{2}$　　Missing wavelengths are 720 nm, 514 nm, and 400 nm.

17* ••　A Newton's-ring apparatus consists of a glass lens with radius of curvature R that rests on a flat glass plate as shown in Figure 35-39. The thin film is air of variable thickness. The pattern is viewed by reflected light. (a) Show that for a thickness t the condition for a bright (constructive) interference ring is

$$t = \left(m + \frac{1}{2} \right) \frac{\lambda}{2}, \quad m = 0, 1, 2, \dots$$

(b) Apply the Pythagorean theorem to the triangle of sides r, $R - t$, and hypotenuse R to show that for $t \ll R$, the radius of a fringe is related to t by

$$r = \sqrt{2tR}$$

(c) How would the transmitted pattern look in comparison with the reflected one? (d) Use $R = 10$ m and a diameter of 4 cm for the lens. How many bright fringes would you see if the apparatus were illuminated by yellow sodium light ($\lambda \approx 590$ nm) and were viewed by reflection? (e) What would be the diameter of the sixth bright fringe? (f) If the glass used in the apparatus has an index of refraction $n = 1.5$ and water ($n_w = 1.33$) is placed between the two pieces of glass, what change will take place in the bright fringes?

(a) This arrangement is essentially identical to the "thin film" configuration, except that the "film" is air. Now the phase change of 180° occurs at the lower reflection. So the condition for constructive interference is $2t/\lambda = m + \frac{1}{2}$ or $t = (m + \frac{1}{2})\lambda/2$. Note that the first bright fringe corresponds to $m = 0$.

(b) From Figure 35-39 we have $r^2 + (R - t)^2 = R^2 = r^2 + R^2 - 2Rt + t^2$. For $t \ll R$ we neglect the last term; solving for r one finds $r = \sqrt{2Rt}$.

(c) As discussed in Problem 9, the transmitted pattern is complimentary to the reflected pattern.

(d) From (a) and (b), $r^2 = (m + \frac{1}{2})\lambda R$; solve for m　　$m = 67$; there will be 68 bright fringes

(e) $D = 2\sqrt{(m + \frac{1}{2})\lambda R}$; solve for $m = 5$　　$D = 1.14$ cm

(f) Now λ in the film is $\lambda_{air}/n = 444$ nm. So the separation between fringes is reduced and the number of fringes that will be seen is increased by the factor $n = 1.33$.

21* •　Two narrow slits separated by 1 mm are illuminated by light of wavelength 600 nm, and the interference pattern is viewed on a screen 2 m away. Calculate the number of bright fringes per centimeter on the screen.

1. From Equ. 35-5, $\Delta y = \lambda L/d$; $N = 1/\Delta y$　　　　　$\Delta y = 1.2$ mm $= 0.12$ cm; $N = 8.33$/cm

---

25* ••　Light is incident at an angle ϕ with the normal to a vertical plane containing two slits of separation d (Figure 35-40). Show that the interference maxima are located at angles θ given by $\sin \theta + \sin \phi = m\lambda/d$.

Note that the total path difference is $d \sin \phi + d \sin \theta$. For constructive interference, $d \sin \phi + d \sin \theta = m\lambda$. Thus, $\sin \phi + \sin \theta = m\lambda/d$ is the condition for interference maxima.

29* •　Equation 35-2, $d \sin \theta = m\lambda$, and Equation 35-11, $a \sin \theta = m\lambda$, are sometimes confused. For each equation, define the symbols and explain the equation's application.

Equ. 35-2 expresses the condition for an intensity maximum in two-slit interference. Here d is the slit separation, λ the wavelength of the light, m an integer, and θ the angle at which the interference maximum appears.

Equ. 35-11 expresses the condition for the first minimum in single-slit diffraction. Here a is the width of the slit, λ the wavelength of the light, and θ the angle at which the first minimum appears.

33★ ••   For a ruby laser of wavelength 694 nm, the end of the ruby crystal is the aperture that determines the diameter of the light beam emitted. If the diameter is 2 cm and the laser is aimed at the moon, 380,000 km away, find the approximate diameter of the light beam when it reaches the moon, assuming the spread is due solely to diffraction.

Use Equ. 35-25; then diameter at moon = θR_{EM} $\theta = 4.23 \times 10^{-5}$ rad; $d = 2 \times 3.8 \times 10^{8} \times 4.23 \times 10^{-5}$ m = 32.2 km

---

37★ ••   Suppose that the *central* diffraction maximum for two slits contains 17 interference fringes for some wavelength of light. How many interference fringes would you expect in the first *secondary* diffraction maximum? There are 8 interference fringes on each side of the central maximum. The secondary diffraction diffraction maximum is half as wide as the central one. It follows that it will contain 8 interference maxima.

41★ ••   At the second secondary maximum of the diffraction pattern of a single slit, the phase difference between the waves from the top and bottom of the slit is approximately 5π. The phasors used to calculate the amplitude at this point complete 2.5 circles. If I_0 is the intensity at the central maximum, find the intensity I at this second secondary maximum.

1. Let A_0 be the amplitude at central maximum $I_0 = CA_0^2$; C is a constant

2. Find A such that $(5\pi/2)A = A_0$; $I = CA^2$ $A = 2A_0/5\pi$, $I = (4/25\pi^2)I_0 = 0.0162I_0$

45★ •••   Three slits, each separated from its neighbor by 0.06 mm, are illuminated by a coherent light source of wavelength 550 nm. The slits are extremely narrow. A screen is located 2.5 m from the slits. The intensity on the centerline is 0.05 W/m². Consider a location 1.72 cm from the centerline. (*a*) Draw the phasors, according to the phasor model for the addition of harmonic waves, appropriate for this location. (*b*) From the phasor diagram, calculate the intensity of light at this location.

(*a*) 1. Determine δ for adjacent slits   From Equs. 35-4 and 35-5, $\delta = 2\pi dy/\lambda L = 3\pi/2$ rad

2. The three phasors, 270° apart, are shown in the diagram. Note that they form three sides of a square. Consequently, their sum, here shown as the resultant R, equals the magnitude of one of the phasors.

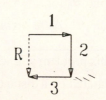

(*b*) Note that $I \propto R^2$ and $I_0 \propto 9R^2$ $I = I_0/9 = 0.00556$ W/m²

49★ •   Two sources of light of wavelength 700 nm are 10 m away from the pinhole of Problem 48. How far apart must the sources be for their diffraction patterns to be resolved by Rayleigh's criterion?

Use Equ. 35-26 to find α_c; then $\Delta y = L\alpha_c$ $\alpha_c = 8.54 \times 10^{-3}$ rad; $\Delta y = 8.54$ cm

---

53★ ••   (*a*) How far apart must two objects be on the moon to be resolved by the eye? Take the diameter of the pupil of the eye to be 5 mm, the wavelength of the light to be 600 nm, and the distance to the moon to be 380,000 km. (*b*) How far apart must the objects on the moon be to be resolved by a telescope that has a mirror of diameter 5 m?

(*a*) Proceed as in Problem 49 $\alpha_c = 1.46 \times 10^{-4}$ rad; $\Delta y = 55.6$ km

(b) Repeat with $D = 5$ m $\alpha_c = 1.46 \times 10^{-7}$ rad; $\Delta y = 55.6$ m

---

57* •   When a diffraction grating is illuminated with white light, the first-order maximum of green light

(a) is closer to the central maximum than that of red light.

(b) is closer to the central maximum than that of blue light.

(c) overlaps the second order maximum of red light.

(d) overlaps the second order maximum of blue light.

(a)

61* •   What is the longest wavelength that can be observed in the fifth-order spectrum using a diffraction grating with 4000 slits per centimeter?

Find largest λ such that $5\lambda/d = \sin\theta = 1$ $d = (1/4000)$ cm; $\lambda = d/5 = 500$ nm

---

65* ••   Sodium light of wavelength 589 nm falls normally on a 2-cm-square diffraction grating ruled with 4000 lines per centimeter. The Fraunhofer diffraction pattern is projected onto a screen at 1.5 m by a lens of focal length 1.5 m placed immediately in front of the grating. Find (a) the positions of the first two intensity maxima on one side of the central maximum, (b) the width of the central maximum, and (c) the resolution in the first order.

(a) From Equ. 35.3, $y_m = m\lambda L/d$; $d = 1/n$ $d = 2.5 \times 10^{-6}$ m; $y_1 = 0.353$ m; $y_2 = 0.707$ m

(b) $\theta_{min} = \lambda/Nd$ (see p. 1130); $\Delta y = 2\theta_{min}L$ $\theta_{min} = 2.95 \times 10^{-5}$ rad; $\Delta y = 88.4$ μm

(c) Use Equ. 35-27 with $m = 1$ $R = N = 8000$

69* ••   White light is incident normally on a transmission grating and the spectrum is observed on a screen 8.0 m from the grating. In the second-order spectrum, the separation between light of 520- and 590-nm wavelength is 8.4 cm.

(a) Determine the number of lines per centimeter of the grating. (b) What is the separation between these two wavelengths in the first-order and third-order spectra?

We will assume that the angle θ_2 is small and then verify that this is a justified assumption.

(a) 1. From Equ. 35-3, $y_2 - y_1 = mL(\lambda_2 - \lambda_1)/d$ $d = mL(\lambda_2 - \lambda_1)/(y_2 - y_1) = 1.333 \times 10^{-5}$ m

   2. $n = 1/d$ $n = 750$ lines/cm

   3. Find θ_2 for $\lambda = 590$ nm $\theta_2 = \sin^{-1}(2\lambda/d) = 8.85 \times 10^{-2} \ll 1$; $\sin\theta \approx \theta$

(b) For $m = 1$, $\Delta y = \Delta y(m = 2)/2$ $m = 1$, $\Delta y = 4.2$ cm; $m = 3$, $\Delta y = 3 \times 4.2$ cm $= 12.6$ cm

---

73* •••   In this problem you will derive Equation 35-28 for the resolving power of a diffraction grating containing N slits separated by a distance d. To do this you will calculate the angular separation between the maximum and minimum for some wavelength λ and set it equal to the angular separation of the mth-order maximum for two nearby wavelengths. (a) Show that the phase difference ϕ between the light from two adjacent slits is given by

$$\phi = \frac{2\pi d}{\lambda} \sin\theta$$

(b) Differentiate this expression to show that a small change in angle $d\theta$ results in a change in phase of $d\phi$ given by

$$d\phi = \frac{2\pi d}{\lambda} \cos\theta \, d\theta$$

(c) For N slits, the angular separation between an interference maximum and interference minimum corresponds to a phase change of $d\phi = 2\pi/N$. Use this to show that the angular separation $d\theta$ between the maximum and minimum for some wavelength λ is given by

$$d\theta = \frac{\lambda}{Nd\cos\theta}$$

(d) The angle of the mth-order interference maximum for wavelength λ is given by Equation 35-27. Compute the differential of each side of this equation to show that angular separation of the mth-order maximum for two nearly equal wavelengths differing by $d\lambda$ is given by

$$d\theta \approx \frac{m\,d\lambda}{d\cos\theta}$$

(e) According to Rayleigh's criterion, two wavelengths will be resolved in the mth order if the angular separation of the wavelengths given by Equation 35-31 equals the angular separation of the interference maximum and interference minimum given by Equation 35-30. Use this to derive Equation 35-28 for the resolving power of a grating.

(a) The path difference for two adjacent slits for an angle θ is $\Delta = d\sin\theta$. The phase difference is $\phi = 2\pi\Delta/\lambda = (2\pi d/\lambda)\sin\theta$.

(b) $d\phi/d\theta = (2\pi d/\lambda)\cos\theta$; so $d\phi = (2\pi d/\lambda)\cos\theta\,d\theta$.

(c) From (b), with $d\phi = 2\pi/N$ we have $d\theta = \lambda/(Nd\cos\theta)$.

(d) $m\lambda = d\sin\theta$; differentiate with respect to λ. $m = d\cos\theta\,(d\theta/d\lambda)$ and $d\theta = (m\,d\lambda)/(d\cos\theta)$.

(e) We now have two expressions for $d\theta$. Equating these gives $\lambda/d\lambda = R = mN$.

77*•   In a lecture demonstration, a laser beam of wavelength 700 nm passes through a vertical slit 0.5 mm wide and hits a screen 6 m away. Find the horizontal length of the principal diffraction maximum on the screen; that is, find the distance between the first minimum on the left and the first minimum on the right of the central maximum.

Use Equ. 35-12; width $= 2y = 2\lambda L/a$ $\qquad\qquad$ $2y = 1.68$ cm

---

81*••   A *Fabry–Perot interferometer* consists of two parallel, half-silvered mirrors separated by a small distance a. Show that when light is incident on the interferometer with an angle of incidence θ, the transmitted light will have maximum intensity when $a = m\lambda/2\cos\theta$.

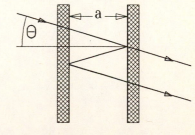

The *Fabry-Perot interfereometer* is shown in the figure. The path difference between the two rays that emerge from the interferometer is $\Delta r = 2a/\cos\theta$. For constructive interference we require $\Delta r = m\lambda$. It follows that the intensity will be a maximum when $a = (m\lambda/2)\cos\theta$.

85*••   The Impressionist painter Georges Seurat used a technique called "pointillism," in which his paintings are composed of small, closely spaced dots of pure color, each about 2 mm in diameter. The illusion of the colors blending together smoothly is produced in the eye of the viewer by diffraction effects. Calculate the minimum viewing distance for this effect to work properly. Use the wavelength of visible light that requires the *greatest* distance, so that you're sure the effect will work for *all* visible wavelengths. Assume the pupil of the eye has a diameter of 3 mm.

1. Write the angle subtended by adjacent dots in terms $\theta = d/L$
of the viewing distance L and dot separation d

2. Set $\theta = \alpha_c$ for shortest λ ($\lambda = 400$ nm); solve for L $L = Dd/1.22\lambda = \dfrac{3\times10^{-3}\times2\times10^{-3}}{1.22\times4\times10^{-9}}$ m $= 12.3$ m

---

89* ••• For the situation described in Problem 88, show that the rate of oscillation of the picture's intensity is a minimum when the airplane is directly above the midpoint between the transmitter and receiving antenna.

There are two paths from the transmitter, T, to the receiver, R, the direct path of length L and the path from T to P, the plane, and from P to R. The path difference is $\Delta r = \sqrt{x^2 + H^2} + \sqrt{(L-x)^2 + H^2} - L$. The rate of oscillation, assuming the plane travels at constant speed, will be a minimum when $d(\Delta r)/dx$ is a minimum. We set the derivative equal to zero and solve for x.

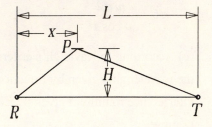

$$\frac{d(\Delta r)}{dx} = \frac{x}{\sqrt{x^2 + H^2}} - \frac{L-x}{\sqrt{(L-x)^2 + H^2}} = 0; \ x = L/2.$$ The rate of oscillation is a minimum when the plane is

midway between the transmitter and receiver.

CHAPTER 36

Applications of the Schrödinger Equation

1* • True or false: Boundary conditions on the wave function lead to energy quantization.

True

5* •• Use the procedure of Example 36-1 to verify that the energy of the first excited state of the harmonic oscillator is $E_1 = \frac{3}{2}\hbar\omega_0$. (*Note*: Rather than solve for a again, use the result $a = m\omega_0/2\hbar$ obtained in Example 36-1.)

The wave function is $\psi = A_1 x e^{-ax^2}$ (see Equ. 36-25). Then $\dfrac{d\psi}{dx} = A_1 e^{-ax^2} - 2ax^2 A_1 e^{-ax^2}$ and

$\dfrac{d^2\psi}{dx^2} = -2axA_1 e^{-ax^2} - 4axA_1 e^{-ax^2} + 4a^2x^3 A_1 e^{-ax^2} = (4a^2x^3 - 6ax)A_1 e^{-ax^2}$. We now substitute this into the

Schrödinger equation. The exponentials and the constant A_1 cancel, so $-(\hbar^2/2m)(4a^2x^3 - 6ax) + \frac{1}{2}m\omega_0^2 x^3 = E_1 x$. With $a = \frac{1}{2}m\omega_0/\hbar$, the terms in x^3 cancel and solving for the energy E_1 we find $E_1 = 6\hbar^2 a/2m = 3\hbar\omega_0/2 = 3E_0$.

9* ••• Verify that $\psi_1(x) = A_1 x e^{-ax^2}$ is the wave function corresponding to the first excited state of a harmonic oscillator by substituting it into the time-independent Schrödinger equation and solving for a and E.

From Problem 5 we know that the Schrödinger equation for ψ_1 gives $-(\hbar^2/2m)(4a^2x^3 - 6ax) + \frac{1}{2}m\omega_0^2 x^3 = E_1 x$. If we now set the coefficients of $x^3 = 0$ and solve for a we find that $a = \frac{1}{2}m\omega_0/\hbar$, and using this expression and solving for the energy E_1 we find $E_1 = 6\hbar^2 a/2m = 3\hbar\omega_0/2 = 3E_0$.

13* •• A free particle of mass m with wave number k_1 is traveling to the right. At $x = 0$, the potential jumps from zero to U_0 and remains at this value for positive x. (*a*) If the total energy is $E = \hbar^2 k_1^2/2m = 2U_0$, what is the wave number k_2 in the region $x > 0$? Express your answer in terms of k_1 and in terms of U_0. (*b*) Calculate the reflection coefficient R at the potential step. (*c*) What is the transmission coefficient T? (*d*) If one million particles with wave number k_1 are incident upon the potential step, how many particles are expected to continue along in the positive x direction? How does this compare with the classical prediction?

(*a*) We are given that $E = \hbar^2 k_1^2/2m = 2U_0$. For $x > 0$, $\hbar^2 k_2^2/2m + U_0 = 2U_0$. So $k_2 = \sqrt{2mU_0}/\hbar$, whereas $k_1 = \sqrt{4mU_0}/\hbar$. $k_2 = k_1/\sqrt{2}$.

(*b*) The reflection coefficient is given by Equ. 36-27. So $R = \dfrac{(k_1 - k_2)^2}{(k_1 + k_2)^2} = 0.0294$.

(c) $T = 1 - R = 0.971$.

(d) The number of particles that continue beyond the step is $N_0 T = 9.71 \times 10^5$; classically, 1×10^6 would continue to move past the step.

17* ••   Use Equation 36-29 to calculate the order of magnitude of the probability that a proton will tunnel out of a nucleus in one collision with the nuclear barrier if it has energy 6 MeV below the top of the potential barrier and the barrier thickness is 10^{-15} m.

1. Rewrite α of Equ. 36-29 in units of MeV $\alpha = \sqrt{2mc^2(U_0 - E)}/\hbar c$; $\hbar c = 1.974 \times 10^{-13}$ MeV·m

2. $T = e^{-2\alpha a}$; $m_p c^2 = 938$ MeV $T = \exp[-2 \times 10^{-15}(2 \times 938 \times 6)^{\frac{1}{2}}/1.974 \times 10^{-13}] = 0.341$

21* •   (a) Repeat Problem 19 for the case $L_2 = 2L_1$ and $L_3 = 4L_1$. (b) What quantum numbers correspond to degenerate energy levels?

(a) From Equ. 36-32, $E = (h^2/8mL_1^2)(n_1^2 + n_2^2/4 + n_3^2/16) = (h^2/128mL_1^2)(16n_1^2 + 4n_2^2 + n_3^2)$. The table below lists the ten lowest energy levels in units of $h^2/128mL_1^2$.

n_1	n_2	n_3	E
1	1	1	21
1	1	2	24
1	1	3	29
1	2	1	33
1	1	4	36
1	2	2	36
1	2	3	41
1	1	5	45
1	2	4	48
1	3	1	53
1	1	6	56
1	3	2	56

(b) There are two degenerate levels, namely, (1,1,4) and (1,2,2) and (1,1,6) and (1,3,2).

25* ••   What is the next energy level above those found in Problem 24c for a particle in a two-dimensional square box for which the degeneracy is greater than 2?

We need to find the least integral values for n and m such that $n^2 + m^2$ are the same for more than two choices of n and m. For any pair of values, e.g, $n = 1$, $m = 2$, and $n = 2$, $m = 1$ we have double degeneracy. Therefore, we must find two different sets for which the sum of the squares are the same. For $n = 1$, $m = 7$ and for $n = 5$, $m = 5$, the sum of the squares equals 50. Consequently, the states $n = 1$, $m = 7$; $n = 7$, $m = 1$; and $n = 5$, $m = 5$ are degenerate (triple degeneracy). The next higher degeneracy is for $n = 4$, $m = 7$; $n = 7$, $m = 4$; $n = 1$, $m = 8$; and $n = 8$, $m = 1$. These states are four-fold degenerate.

29* ••   Show that the ground-state wave function and that of the first excited state of the harmonic oscillator are orthogonal; i.e., show that $\int \psi_0(x)\psi_1(x)\, dx = 0$.

We need to show that $\int_{-\infty}^{\infty} \psi_0(x)\,\psi_1(x)\,dx = 0$, where $\psi_0(x)$ and $\psi_1(x)$ are given by Equs. 36-23 and 36-25,

respectively. Note that $\psi_0(x)$ is an even function of x and $\psi_1(x)$ is an odd function of x. It follows that the integral from $-\infty$ to ∞ must vanish.

33* •• Eight identical noninteracting fermions (such as neutrons) are confined to a two-dimensional square box of side length L. Determine the energies of the three lowest states. (See Problem 26.)

Each n, m state can accommodate only 2 particles. Therefore, in the ground state of the system of 8 fermions, the four lowest quantum states are occupied. These are (1,1), (1,2), (2,1) and (2,2). [Note that the states (1,2) and (2,1) are distinctly different states since the x and y directions are distinguishable.] The energy of the ground state is $E_0 = 2(h^2/8mL^2)(2 + 5 + 5 + 8) = 5h^2/mL^2$. The next higher state is achieved by taking one fermion from the (2,2) state and raising it into the next higher unoccupied state. That state is the (1,3) state. The energy difference between the ground state and this state is $(h^2/8mL^2)(10 - 8) = h^2/4mL^2$. The (3,1) is another excited state that is accessible, and it is degenerate with the (1,3) state. The three lowest energy levels are therefore $E_0 = 5h^2/mL^2$, and two states of energy $E_1 = E_2 = 21h^2/4mL^2$.

37* ••• Show that Equations 36-27 and 36-28 imply that the transmission coefficient for particles of energy E incident on a step barrier $U_0 < E$ is given by

$$T = \frac{4k_1 k_2}{(k_1 + k_2)^2} = \frac{4r}{(1 + r)^2}$$

where $r = k_2/k_1$.

$$R = \frac{(k_1 - k_2)^2}{(k_1 + k_2)^2} \quad \text{and} \quad T = 1 - R = \frac{(k_1 + k_2)^2 - (k_1 - k_2)^2}{(k_1 + k_2)^2} = \frac{4k_1 k_2}{(k_1 + k_2)^2} = \frac{4r}{(1 + r)^2} \quad \text{where } r = k_2/k_1.$$

41*••• In this problem you will derive the ground-state energy of the harmonic oscillator using the precise form of the uncertainty principle, $\Delta x\,\Delta p \geq \hbar/2$, where Δx and Δp are defined to be the standard deviations $(\Delta x)^2 = [(x - x_{av})^2]_{av}$ and $(\Delta p)^2 = [(p - p_{av})^2]_{av}$ (see Equation 18-31). Proceed as follows:

1. Write the total classical energy in terms of the position x and momentum p using $U(x) = \frac{1}{2}m\omega^2 x^2$ and $K = p^2/2m$.

2. Use the result of Equation 18-35 to write $(\Delta x)^2 = [(x - x_{av})^2]_{av} = (x^2)_{av} - x_{av}^2$ and $(\Delta p)^2 = [(p - p_{av})^2]_{av} = (p^2)_{av} - p_{av}^2$.

3. Use the symmetry of the potential energy function to argue that x_{av} and p_{av} must be zero, so that $(\Delta x)^2 = (x)_{av}$ and $(\Delta p)^2 = (p^2)_{av}$.

4. Assume that $\Delta p = \hbar/2\Delta x$ to eliminate $(p^2)_{av}$ from the average energy $E_{av} = (p^2)_{av}/2m + \frac{1}{2}m\omega^2(x^2)_{av}$ and write E_{av} as $E_{av} = \hbar^2/8mZ + \frac{1}{2}m\omega^2 Z$, where $Z = (x^2)_{av}$.

5. Set $dE/dZ = 0$ to find the value of Z for which E is a minimum.

6. Show that the minimum energy is given by $(E_{av})_{min} = +\frac{1}{2}\hbar\omega$.

1. $E_{av} = U_{av} + K_{av} = \frac{1}{2}m\omega^2(x^2)_{av} + (1/2m)(p^2)_{av}$.

2, 3. $(\Delta p)^2 = [(p - p_{av})^2]_{av} = [p^2 - 2pp_{av} + p_{av}^2]_{av} = (p^2)_{av}$ since $p_{av} = 0$; likewise, $(\Delta x)^2 = (x^2)_{av}$.

4. $E_{av} = \frac{1}{2}m\omega^2(x^2)_{av} + (\hbar^2/8m)/(x^2)_{av} = \frac{1}{2}m\omega^2 Z + \hbar^2/8mZ$.

5. $dE_{av}/dZ = \frac{1}{2}m\omega^2 - \hbar^2/8mZ^2 = 0$; $Z = \hbar/2m\omega$.

6. $(E_{av})_{min} = \frac{1}{2}m\omega^2\hbar/2m\omega + 2m\omega\hbar^2/8m\omega = \frac{1}{2}\hbar\omega$.

CHAPTER **37**

Atoms

1* • As n increases, does the spacing of adjacent energy levels increase or decrease?

The spacing decreases.

5* •• The kinetic energy of the electron in the ground state of hydrogen is 13.6 eV = E_0. The kinetic energy of the electron in the state $n = 2$ is _____.

(*a*) $4E_0$, (*b*) $2E_0$, (*c*) $E_0/2$, (*d*) $E_0/4$.

(*d*)

9* • Find the photon energy for the three longest wavelengths in the Balmer series and calculate the wavelengths.

For the Balmer series, $E_f = E(n = 2) = -3.40$ eV. Use Equs. 17.5 and 17.21, i.e., $\lambda = (1240 \text{ eV·nm})/(\Delta E \text{ eV})$.

1. $\Delta E = E_3 - E_2$; $E_3 = -13.6/9$ eV $= -1.51$ eV $\Delta E = 1.89$ eV; $\lambda_{3,2} = 656.1$ nm

2. $\Delta E = E_4 - E_2$; $E_4 = -13.6/16$ eV $= -0.85$ eV $\Delta E = 2.55$ eV; $\lambda_{4,2} = 486.3$ nm

3. $\Delta E = E_5 - E_2$; $E_5 = -13.6/25$ eV $= -0.544$ eV $\Delta E = 2.856$ eV; $\lambda_{5,2} = 434.2$ nm

---

13* •• The binding energy of an electron is the minimum energy required to remove the electron from its ground state to a large distance from the nucleus. (*a*) What is the binding energy for the hydrogen atom? (*b*) What is the binding energy for He$^+$? (*c*) What is the binding energy for Li^{2+}? [Singly ionized helium (He$^+$) and doubly ionized lithium (Li^{2+}) are "hydrogen-like" in that the system consists of a positively charged nucleus and a single bound electron.]

(*a*) $BE = E_\infty - E(n = 1) = E_0$ $BE = 13.6$ eV

(*b*) Note that $E_n \propto Z^2$; Z for He$^+$ = 2 $BE = 4 \times 13.6$ eV $= 54.4$ eV

(*c*) For Li^{2+}, $Z = 3$ $BE = 9 \times 13.6$ eV $= 122.4$ eV

17* •• In a reference frame with the origin at the center of mass of an electron and the nucleus of an atom, the electron and nucleus have equal and opposite momenta of magnitude p. (*a*) Show that the total kinetic energy of the electron and nucleus can be written $K = p^2/2m_r$, where $m_r = m_e M/(M + m_e)$ is called the reduced mass, m_e is the mass of the electron, and M is the mass of the nucleus. It can be shown that the motion of the nucleus can be accounted for by replacing the mass of the electron by the reduced mass. In general, the reduced mass for a two-body problem with masses m_1 and m_2 is given by

$$m_r = \frac{m_1 m_2}{m_1 + m_2}$$ (Eq. 37-47)

(b) Use Equation 37-14 with m replaced by m_r to calculate the Rydberg constant for hydrogen ($M = m_p$) and for a very massive nucleus ($M = \infty$). (c) Find the percentage correction for the ground-state energy of the hydrogen atom due to the motion of the proton.

(a) Both the electron and nucleus have the same momentum, so $K = K_e + K_n = p^2/2m_e + p^2/2M = (p^2/2)[(m_e + M/m_e M)]$. Defining $m_r = m_e M/(m_e + M)$, $K = p^2/2m_r$.

(b) For H, $R_H = 1.096776 \times 10^7$ m^{-1} $= Cm_r$, where $C = k^2 e^4/4\pi c\hbar^3$, and $m_r = m_p m_e/(m_e + m_p) = m_e/(1 + m_e/m_p) = m_e/(1 + 5.447 \times 10^{-4})$. If instead of m_p we have an infinite nuclear mass, then $m_r = m_e$, and the Rydberg constant R_∞ is then greater than R_H by the factor $(1 + 5.447 \times 10^{-4})$, or $R_\infty = 1.097373 \times 10^7$ m^{-1}.

(c) The energy correction is just -5.447×10^{-4} or -0.0545%. That is, the correct energy is slightly less than that calculated neglecting the motion of the nucleus.

21**•   Work Problem 20 for $\ell = 3$.

(a) Use Equ. 37-24 $L = 2\sqrt{3}\hbar$

(b) $m = -\ell \dots 0 \dots +\ell$ $m = -3, -2, -1, 0, +1, +2, +3$

(c) The vector diagram is shown on the right. Note that since $L_z = m\hbar$ and $L = 2\sqrt{3}\hbar$, the angles between the vectors and the z axis are given by $\cos\theta_m = m/2\sqrt{3}$. Thus, $\theta_3 = 30°$, $\theta_2 = 54.7°$, and $\theta_1 = 73.2°$. As shown, the spacing between the allowed values of L_z is constant and equal to $\hbar$.

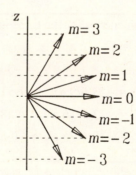

25*•   Find the minimum value of the angle θ between L and the z axis for (a) $\ell = 1$, (b) $\ell = 4$, and (c) $\ell = 50$. We consider the general case. The minimum angle between the z axis and L is the angle between the L vector for $m = \ell$ and the z axis. In this case, $L_z = m\hbar = \ell\hbar$ and $L = \sqrt{\ell(\ell + 1)}\hbar$. Consequently, $\cos\theta = \sqrt{\ell/(\ell + 1)}$.

(a) For $\ell = 1$, $\cos\theta = 1/\sqrt{2}$ $\theta = 45°$

(b) For $\ell = 4$, $\cos\theta = 2/\sqrt{5}$ $\theta = 26.6°$

(c) For $\ell = 50$, $\cos\theta = \sqrt{50/51}$ $\theta = 8.05°$

29*•   (a) If spin is not included, how many different wave functions are there corresponding to the first excited energy level $n = 2$ for hydrogen? (b) List these functions by giving the quantum numbers for each state.

(a), (b) For $n = 2$, $\ell = 0$ or 1; for $\ell = 1$, there are three values of m, namely $m = -1, 0, 1$

There are four functions; they are listed by (n, ℓ, m):

$(2,0,0),\ (2,1,-1),\ (2,1,0),\ (2,1,1)$

33*•• Calculate the probability of finding the electron in the range $\Delta r = 0.02a_0$ at (a) $r = a_0$ and (b) $r = 2a_0$ for the

state $n = 2$, $\ell = 0$, $m = 0$ in hydrogen. (See Problem 31 for the value of $C_{2,0,0}$.)

In this instance, $\int P(r)\, dr$ extends over a sufficiently narrow interval $\Delta r \ll a_0$ that one may neglect the dependence

of $P(r)$ on r, i.e., we will set $\int P(r)\, dr = P(r)\, \Delta r$.

(a) 1. Use Equ. 37-36 and $C_{2,0,0}$ as given; $Z = 1$ $\qquad \psi_{2,0,0} = \dfrac{1}{4\sqrt{4\pi}}\left(\dfrac{1}{a_0}\right)^{3/2}\left(2 - \dfrac{r}{a_0}\right)e^{-r/2a_0}$.

2. Set $r = a_0$ and evaluate $\qquad\qquad\qquad\qquad \psi_{2,0,0}(a_0) = 0.0605/a_0^{3/2}$

3. Square the result of (1) $\qquad\qquad\qquad\qquad (\psi_{2,0,0}(a_0))^2 = 0.00366/a_0^3$

4. $P(r) = 4\pi r^2 \psi^2(r)$, Equ. 37-34 $\qquad\qquad P(a_0) = 0.046/a_0$

5. Probability $= P(a_0)\Delta r$ $\qquad\qquad\qquad$ Probability $= 0.046 \times 0.02 = 9.2 \times 10^{-4}$

(b) Use (a) part 1; find $\psi_{2,0,0}(2a_0)$ and $P(2a_0)$ $\qquad P(a_0) = 0$; Probability $= 0$

37*••• Calculate the probability that the electron in the ground state of a hydrogen atom is in the region

$0 < r < a_0$.

We evaluate $\displaystyle\int_0^{a_0} 4\pi r^2\, \psi_{1,0,0}^2(r)\, dr = (4/a_0^3)\int_0^{a_0} r^2 e^{-2r/a_0}\, dr = -e^{-2r/a_0}(2r^2/a_0^2 + 2r/a_0 + 1)\Big|_0^{a_0} = 1 - 5e^{-2} = 0.323$,

where we have used $\displaystyle\int x^2 e^{bx}\, dx = (e^{bx}/b^3)(b^2 x^2 - 2bx + 2)$.

41*• A hydrogen atom is in the state $n = 3$, $\ell = 2$. (a) What are the possible values of j?

$j = \ell \pm \tfrac{1}{2}$ $\qquad\qquad\qquad\qquad\qquad\qquad j = 3/2$ or $5/2$

45*•• The properties of iron ($Z = 26$) and cobalt ($Z = 27$), which have adjacent atomic numbers, are

similar, whereas the properties of neon ($Z = 10$) and sodium ($Z = 11$), which also have adjacent atomic

numbers, are very different. Explain why.

For iron and cobalt, the closed shell configuration is the same; the only difference is the number of $3d$ electrons.

Neon is a closed shell configuration ($n = 1, 2$), whereas sodium has one extra electron outside a closed shell. The

properties of neon are therefore similar to those of helium and other rare gas atoms, whereas those of sodium are

similar to those of hydrogen and other atoms (lithium, potassium) with one electron outside a closed shell.

49*• How many of oxygen's eight electrons are found in the p state? (a) 0, (b) 2, (c) 4, (d) 6, (e) 8.

(c)

53*•• If the outer electron in sodium moves in the $n = 3$ Bohr orbit, the effective nuclear charge would be $Z'e =$

$1e$, and the energy of the electron would be -13.6 eV/$3^2 = -1.51$ eV. However, the ionization energy of sodium

is 5.14 eV, not 1.51 eV. Use this fact and Equation 37-45 to calculate the effective nuclear charge Z' seen by the

outer electron in sodium. Assume that $r = 9a_0$ for the outer electron.

Use Equ. 37-45, assuming $r = 9a_0$ $\qquad\qquad Z' = 9(5.14/13.6) = 3.40$

[Note: If we use the Bohr model: $E_n = -(Z')^2 E_0/n^2$] $\qquad [Z' = \sqrt{9(5.14/13.6)} = 1.84]$

[The above result indicates that the assumption that $r = 9a_0$ is not justified.]

57*• (a) Calculate the next two longest wavelengths in the K series (after the K_α line) of molybdenum. (b) What is

the wavelength of the shortest wavelength in this series?

(a) Use Equ. 37-46 with $n = 3$ and $Z = 42$; repeat for $n = 4$

$\lambda = [1240/(41^2 \times 13.6 \times 0.889)]$ nm $= 0.0610$ nm

$\lambda = 0.0610(0.889/0.9375)$ nm $= 0.0578$ nm

(b) Repeat part a with $n = \infty$

$\lambda = 0.0542$ nm

61*• Calculate the wavelength of the K_α line in (a) magnesium ($Z = 12$) and (b) copper ($Z = 29$).

(a) Use Equ. 37-46 with $n = 2$ and $Z = 12$

$\lambda = [1240/(11^2 \times 13.6 \times 0.75)]$ nm $= 1.00$ nm

(b) Use Equ. 37-46 with $n = 2$ and $Z = 29$

$\lambda = [1240/(28^2 \times 13.6 \times 0.75)]$ nm $= 0.155$ nm

---

65*•• In Figure 37-17, there are small dips in the ionization-energy curve at $Z = 31$ (gallium) and $Z = 49$ (indium) that are not labeled. Explain these dips using the electron configurations of these atoms given in Table 37-1.

Zinc ($Z = 30$) has a configuration of closed subshells, namely $3d^{10}4s^2$; its ionization energy is relatively large. Gallium, has one more electron in the $4p$ state; this electron is less tightly bound and so the ionization energy of gallium is relatively small. Likewise for cadmium ($Z = 48$) with the outer closed subshell configuration of $4d^{10}5s^2$, and indium with one more electron in the $5p$ state.

69*• Spectral lines of the following wavelengths are emitted by singly ionized helium: 164 nm, 230.6 nm, and 541 nm. Identify the transitions that result in these spectral lines.

1. Express E_n for He$^+$; note that $Z = 2$

$E_n = -(4 \times 13.6/n^2)$ eV $= -54.4/n^2$ eV

2. Use Equs. 37-18 and 37-15 to write n_1 and n_2 in terms of λ

$$\frac{1}{n_1^2} - \frac{1}{n_2^2} = \frac{1240}{54.4\,\lambda}$$

3. By trial and error identify the integers n_1 and n_2 for which $\lambda = 164$ nm, 230.6 nm, and 541 nm.

For $\lambda = 164$ nm, $n_2 = 3$, $n_1 = 2$; for $\lambda = 230.6$ nm, $n_2 = 9$, $n_1 = 3$; for $\lambda = 541$ nm, $n_2 = 7$, $n_1 = 4$

73*•• The combination of physical constants $\alpha = e^2k/\hbar c$, where k is the Coulomb constant, is known as the *fine structure constant*. It appears in numerous relations in atomic physics. (a) Show that α is dimensionless. (b) Show that in the Bohr model of hydrogen $v_n = c\alpha/n$, where v_n is the speed of the electron in the stationary state of quantum number n.

(a) Dimension of ke^2 is $[E][L]$, where $[E]$ denotes the dimension of energy. (See Problem 70.) The dimension of $\hbar c$ is $([E][T])([L]/[T]) = [E][L]$. It follows that $\alpha = ke^2/\hbar c$ is dimensionless.

(b) From Equ. 37-9, $v = n\hbar/mr_n$, and using Equ. 37-11 we have $v = ke^2/\hbar n = \alpha c/n$.

77*•• The triton, a nucleus consisting of a proton and two neutrons, is unstable with a fairly long half-life of about 12 years. *Tritium* is the bound state of an electron and a triton. (a) Calculate the Rydberg constant of tritium using the reduced mass as given by Equation 37-47 in Problem 17. (b) Using the result obtained in (a) and in part (a) of Problem 75 determine the wavelength difference between the longest wavelength Balmer lines of tritium and deuterium and between tritium and hydrogen.

(a) $R_T = R_\infty/(1 + m_e/M_T)$

$R_T = R_\infty/[1 + 1/(3 \times 1836)] = 1.097174 \times 10^7$ m^{-1}

(b) 1. $\Delta\lambda_{D,T} = \lambda[(m_e/M_D) - (m_e/M_T)]$

$\Delta\lambda_{D,T} = 656.47[(1/3672) - (1/5508)]$ nm $= 0.0596$ nm

2. $\Delta\lambda_{H,T} = \lambda[(m_e/M_H) - (m_e/M_T)]$

$\Delta\lambda_{H,T} = 656.47[(1/1836) - (1/5508)]$ nm $= 0.238$ nm

CHAPTER 38

Molecular Bonding

1* •   Would you expect the NaCl molecule to be polar or nonpolar?

NaCl is a polar molecule.

5* •   What kind of bonding mechanism would you expect for (a) the N_2 molecule, (b) the KF molecule, (c) Ag atoms in a solid?

(a) N_2 – covalent bonding. (b) KF – ionic bonding. (c) Ag (solid) – metallic bonding.

9* •   The equilibrium separation of the HF molecule is 0.0917 nm and its measured electric dipole moment is 6.40×10^{-30} C·m. What percentage of the bonding is ionic ?

1. Find dipole moment for 100% ionic bonding $p_{100} = er = 1.6 \times 10^{-19} \times 9.17 \times 10^{-11}$ C·m $= 1.47 \times 10^{-29}$ C·m

2. Percent ionic bonding $= 100(p_{meas}/p_{100})$ Ionic bonding $= 43.6\%$

13* ••   Indicate the mean value of r for two vibration levels in the potential-energy curve for a diatomic molecule and show that because of the asymmetry in the curve, r_{av} increases with increasing vibration energy, and therefore solids expand when heated.

We shall assume that $U(r)$ is of the form given in Problem 16 with $n = 6$. The potential energy curve is shown in the figure. The turning points for vibrations of energy E_1 and E_2 are at the values of r where the energies equal $U(r)$. It is apparent that the average value of r depends on the energy, and that $r_{2,av}$ is greater than $r_{1,av}$.

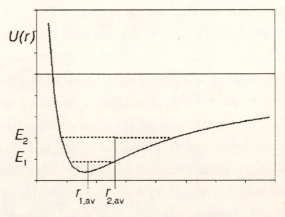

17*•••   (a) Find U_{rep} at $r = r_0$ for NaCl. (b) Assume $U_{rep} = C/r^n$ and find C and n for NaCl. (See Problem 16.)

(a) 1. Find $U_e = -ke^2/r_0$ $U_e = -(1.44/0.236)$ eV $= -6.10$ eV

2. Find $U_{rep} = -(U_e + E_d + \Delta E)$ $U_{rep} = -(-6.10 + 4.27 + 1.52)$ eV $= 0.31$ eV

(b) Using the procedure indicated in Problem 16,

$n = |U_e/U_{rep}|$ and $C = U_{rep} r_0^n$

$n \approx 20; C \approx 8.9 \times 10^{-14}$ eV$\cdot$nm^{-20}

21* •   The characteristic rotational energy E_{0r} for the rotation of the N_2 molecule is 2.48×10^{-4} eV. From this find the separation distance of the N atoms in N_2.

1. From Equ. 38-13, $I = \hbar^2/2E_{0r}$ $I = (1.05 \times 10^{-34}$ J$\cdot$s$)^2/7.94 \times 10^{-23}$ J $= 1.39 \times 10^{-46}$ J$\cdot$s^2

2. $I = 2M_N(r/2)^2 = M_N r^2/2; M_N = 14m_p$ $r = [(2 \times 1.39 \times 10^{-46})/(14 \times 1.67 \times 10^{-27})]^{\frac{1}{2}}$ m $= 0.109$ nm

---

25* ••   Repeat Problem 24 for LiD, where D is the symbol for deuterium. Note that replacing the proton by the deuteron does not change the equilibrium separation between the nuclei of the molecule.

1. Find μ and $E_{0r} = \hbar^2/2I = \hbar^2/2\mu r^2$ $\mu = (2 \times 6.94/8.94)$ u $= 1.55$ u $= 2.58 \times 10^{-27}$ kg;

   Note: $I = \mu r^2$ (see Equ. 38-14) $E_{0r} = (1.05 \times 10^{-34})^2/[5.16 \times 10^{-27}(1.6 \times 10^{-10})^2]$ J

   $= 0.522$ meV

2. Use Equ. 38-12; $\Delta E = 6E_{0r}$ $\Delta E = 3.13$ meV

29* •• Calculate the effective force constant for HCl from its reduced mass and the fundamental vibrational frequency obtained from Figure 38-17.

From Equ. 38-19, $K = 4\pi^2\mu f^2; \mu = 0.973$ u $K = [4\pi^2 \times 0.973 \times 1.66 \times 10^{-27} (8.66 \times 10^{13})^2]$ N/m $= 480$ N/m

---

33* •   Suppose that hard spheres of radius R are located at the corners of a unit cell with a simple cubic structure. (a) If the hard spheres touch so as to take up the minimum volume possible, what is the size of the unit cell? (b) What fraction of the volume of the cubic structure is occupied by the hard spheres?

(a) The cube has a length $a = 2R$ Unit cell is the cube; its volume is $8R^3$

(b) "Packing fraction" = $V_{sphere}/V_{unit cell}$ Fraction = $[(4\pi R^3/3)/(8R^3)] = 0.5236$

37* ••   Suppose identical bowling balls of radius R are packed into a hexagonal close-packed structure. What fraction of the available volume of the unit cell is filled by the bowling balls?

In the adjacent drawing we show the base layer of six spheres (solid circles) and the next layer of spheres (dashed circles) placed in the interstices. From the drawing (see also Fig. 38-20) we see that the three spheres of the lower layer and one of the spheres in the next layer form a regular tetrahedron. The next layer above the spheres shown dashed overlaps the hexagonal layer shown solid. It follows that the distance between the layer of solid spheres and that of the dashed spheres is $a\sqrt{2/3}$, where $a = 2R$. The unit cell is a cylinder whose base is the hexagon formed by the centers of the solid spheres and whose height is the distance between two overlapping layers. The base of the unit cell has an area of $6(a^2\sqrt{3}/4) = 3a^2\sqrt{3}/2$ and the height of the unit cell is $2a\sqrt{2/3}$. Thus the unit cell volume is $3a^3\sqrt{2}$. Each of the six solid spheres on the hexagon is shared by three adjacent unit cells so that these six spheres contribute two spheres to the unit cell. In addition there is the sphere at the center of the base and the three spheres in the

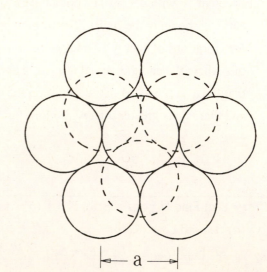

"dashed" layer. Thus the unit cell contains a total of six spheres whose total volume is $6[(4\pi/3)(a^3/8)] = \pi a^3$.

The volume occupied by the spheres is therefore $\pi/(3\sqrt{2})$ of the unit cell volume or 0.74 of the unit cell volume.

---

41 * •   What type of semiconductor is obtained if silicon is doped with (a) indium and (b) antimony? (See Table 37-1 for the electron configurations of these elements.)

(a) p-type (b) n-type

---

45 * •   A doped n-type silicon sample with 10^{16} electrons per cubic centimeter in the conduction band has a resistivity of 5×10^{-3} $\Omega\cdot$m at 300 K. Find the mean free path of the electrons. Use the effective mass of $0.2m_e$ for the mass of the electrons. (See Problem 43.) Compare this mean free path with that of conduction electrons in copper at 300 K.

1. Use Equ. 27-8 to find v_{av}

$$v_{av} = \sqrt{\frac{3(1.38\times10^{-23})(300)}{0.2(9.11\times10^{-31})}} \text{ m/s} = 2.62\times10^5 \text{ m/s}$$

2. Use Equ. 27-7; $\lambda = \dfrac{m_{eff}v_{av}}{\rho n_e e^2}$

$$\lambda = \frac{0.2(9.11^{-31})(2.62\times10^5)}{(5\times10^{-3})(10^{22})(1.6\times10^{-19})^2} \text{ m} = 37.3 \text{ nm}$$

3. $u_F = 1.57\times10^6$ m/s (Example 27-4);
$\rho = 1.7\times10^{-8}$ $\Omega\cdot$m; $n_e = 8.47\times10^{28}$ m^{-3}

$$\lambda_{Cu} = \frac{(9.11\times10^{-31})(1.57\times10^6)}{(1.7\times10^{-8})(8.47\times10^{28})(1.6\times10^{-19})^2} \text{ m} = 38.8 \text{ nm}$$

The mean free paths are about the same

---

49 * ••   Simple theory for the current versus the bias voltage across a pn junction yields the equation

$$I = I_0(e^{eV_b/kT} - 1)$$

Sketch I versus V_b for both positive and negative values of V_b using this equation.

The sketch is shown in the adjoining figure.

Here we plot I/I_0 versus eV/kT where we have

used V to represent the bias voltage V_b.

Except for the region of reverse breakdown,

this curve is identical to Figure 38-27.

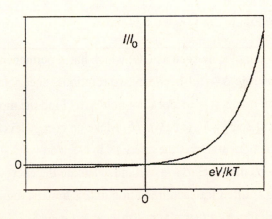

---

53 * ••   Make a sketch showing the valence and conduction band edges and Fermi energy of a pn-junction diode when biased (a) in the forward direction and (b) in the reverse direction.

The sketch of the pn junction biased in the forward (a) and reverse (b) directions is shown below.

The nearly full valence band is shown shaded.

The Fermi level is shown by the dashed line.

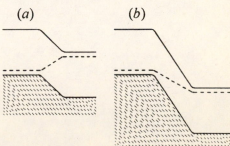

57* •   What kind of bonding mechanism would you expect for (*a*) the HCl molecule (*b*) the O_2 molecule, (*c*) Cu atoms in a solid?

(*a*) Ionic bonding   (*b*) Covalent bonding   (*c*) Metallic bonding

---

61* ••   The equilibrium separation between the nuclei of the CO molecule is 0.113 nm. Determine the energy difference between the $\ell = 2$ and $\ell = 1$ rotational energy levels of this molecule.

1. Find μ and $I = \mu r_0^2$ $\mu = (16 \times 12/28)$ u $= 6.86$ u; $I = 1.45 \times 10^{-46}$ kg·m²

2. $\Delta E_{1,2} = 2\hbar^2/I$ $\Delta E_{1,2} = 1.52 \times 10^{-22}$ J $= 0.948$ meV

65* ••   The resistivity of a sample of pure silicon diminishes drastically when it is irradiated with infrared light of wavelength less than 1.13 μm. What is the energy gap of silicon?

$E_g = hc/\lambda$ $E_g = (1240/1130)$ eV $= 1.10$ eV

69* ••   The potential energy between two atoms in a molecule can often be described rather well by the Lenard-Jones potential, which can be written

$$U = U_0\left[\left(\frac{a}{r}\right)^{12} - 2\left(\frac{a}{r}\right)^6\right]$$

where U_0 and a are constants. Find the interatomic separation r_0 in terms of a for which the potential energy is a minimum. Find the corresponding value of U_{min}. Use Figure 38-4 to obtain numerical values of r_0 and U_0 for the H_2 molecule and express your answers in nanometers and electron volts.

1. Set $dU/dr = 0$ and solve for $r = r_0$ $\dfrac{dU}{dr} = -\dfrac{U_0}{r}\left[12\left(\dfrac{a}{r}\right)^{12} - 12\left(\dfrac{a}{r}\right)^6\right] = 0$; $r_0 = a$

2. From Figure 38-4, $r_0 = 0.074$ nm, $U_{min} = -U_0$ $a = 0.074$ nm; $U_{min} = -4.52$ eV; $U_0 = 4.52$ eV

73* ••   For a molecule such as CO, which has a permanent electric dipole moment, radiative transitions obeying the selection rule $\Delta\ell = \pm 1$ between two rotational energy levels of the same vibrational level are allowed. (That is, the selection rule $\Delta v = \pm 1$ does not hold.) (*a*) Find the moment of inertia of CO and calculate the characteristic rotational energy E_{0r} (in eV). (*b*) Make an energy level diagram for the rotational levels for $\ell = 0$ to $\ell = 5$ for some vibrational level. Label the energies in electron volts starting with $E = 0$ for $\ell = 0$. (*c*) Indicate on your diagram transitions that obey $\Delta\ell = -1$ and calculate the energy of the photon emitted (*d*) Find the wavelength of the photons emitted for each transition in (*c*). In what region of the electromagnetic spectrum are these photons?

(*a*) 1. Find μ and I; $\mu = m_1 m_2/(m_1 + m_2)$; $I = \mu r_0^2$ $\mu = 6.86$ u; $I = (6.86 \times 1.66 \times 10^{-27})(0.113 \times 10^{-9})^2$ kg·m²

       Note: $r_0 = 0.113$ nm $= 1.45 \times 10^{-46}$ kg·m²

       2. $E_{0r} = \hbar^2/2I$ $E_{0r} = [(1.05 \times 10^{-34})^2/2.9 \times 10^{-46}]$ J $= 0.238$ meV

(*b*), (*c*) The energy level diagram is shown in the adjacent figure. Note that $\Delta E_{\ell,\ell-1}$, the energy difference between adjacent levels for $\Delta \ell = -1$, is

$$\Delta E_{\ell,\ell-1} = 2\ell E_{0r}.$$

(*d*) $\lambda_{\ell,\ell-1} = 620/\ell E_{0r}$; λ in nm and E_{0r} in eV. We find:

$\lambda_{1,0} = 2605\ \mu m;\ \lambda_{2,1} = 1303\ \mu m;\ \lambda_{3,2} = 868\ \mu m;$

$\lambda_{4,3} = 651\ \mu m;\ \lambda_{5,4} = 521\ \mu m$

These wavelengths fall in the microwave region of the spectrum.

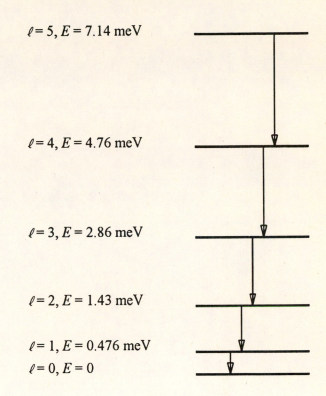

$\ell = 5, E = 7.14$ meV

$\ell = 4, E = 4.76$ meV

$\ell = 3, E = 2.86$ meV

$\ell = 2, E = 1.43$ meV

$\ell = 1, E = 0.476$ meV

$\ell = 0, E = 0$

CHAPTER 39

Relativity

1* • You are standing on a corner and a friend is driving past in an automobile. Both of you note the times when the car passes two different intersections and determine from your watch readings the time that elapses between the two events. Which of you has determined the proper time interval?

By definition, the proper time is measured by the clock in the rest frame of the car, i.e., by the clock in the car.

5* • (a) In the reference frame of the muon in Problem 4, how far does the laboratory travel in a typical lifetime of 2 μs? (b) What is this distance in the laboratory's frame?

(a) $\Delta x = v\Delta t_\mu$ $\Delta x_\mu = 0.999 \times 3 \times 10^8 \times 2 \times 10^{-6}$ m = 599.4 m

(b) $\Delta x' = \gamma\Delta x_\mu$ $\Delta x' = 13.4$ km

9* • A meterstick moves with speed $V = 0.8c$ relative to you in the direction parallel to the stick. (a) Find the length of the stick as measured by you. (b) How long does it take for the stick to pass you?

(a) Use Equ. 39-14 $\gamma = 1/0.6$; $L = (1$ m$)/\gamma = 0.6$ m

(b) $\Delta t = L/V$ $\Delta t = (0.6$ m$)/0.8c = 2.5$ ns

13* •• In the Stanford linear collider, small bundles of electrons and positrons are fired at each other. In the laborator's frame of reference, each bundle is about 1 cm long and 10 μm in diameter. In the collision region, each particle has an energy of 50 GeV, and the electrons and positrons are moving in opposite directions. (a) How long and how wide is each bundle in its own reference frame? (b) What must be the minimum proper length of the accelerator for a bundle to have both its ends simultaneously in the accelerator in its own reference frame? (The actual length of the accelerator is less than 1000 m.) (c) What is the length of a positron bundle in the reference frame of the electron bundle?

(a) 1. Use Equ. 39-25 to find γ $\gamma = 50 \times 10^3/0.511 = 9.785 \times 10^4$

     2. Find proper length of electron bundle; $L_{ep} = \gamma L_e$ $L_{ep} = 978.5$ m; width unchanged = 10 μm

(b) 1. Find length of accelerator in electron frame $L_{acc,e} = L_{acc,p}/\gamma$

     2. Set $L_{acc,e} = L_p$ and solve for $L_{acc,p}$ $L_{acc,p} = (978.5$ m$)\gamma = 9.57 \times 10^7$ m

(c) Find length of positron bundle in electron frame $L_{pos} = (1$ cm$)/\gamma = 1.02 \times 10^{-7}$ m = 0.102 μm

---

17*••   How great must the relative speed of two observers be for the time-interval measurements to differ by 1%? (See Problem 14.)

$(\Delta t - \Delta t')/\Delta t = 1 - 1/\gamma \approx \tfrac{1}{2}V^2/c^2$ $\tfrac{1}{2}V^2/c^2 = 0.01;\ V = \sqrt{0.02}\,c = 0.141c = 4.24\times10^7$ m/s

---

21* •••   Two events in S are separated by a distance $D = x_2 - x_1$ and a time $T = t_2 - t_1$. (a) Use the Lorentz transformation to show that in frame S', which is moving with speed V relative to S, the time separation is $t_2' - t_1' = \gamma(T - VD/c^2)$. (b) Show that the events can be simultaneous in frame S' only if D is greater than cT. (c) If one of the events is the *cause* of the other, the separation D must be less than cT, since D/c is the smallest time that a signal can take to travel from x_1 to x_2 in frame S. Show that if D is less than cT, t_2' is greater than t_1' in all reference frames. This shows that if the cause precedes the effect in one frame, it must precede it in all reference frames. (d) Suppose that a signal could be sent with speed $c' > c$ so that in frame S the cause precedes the effect by the time $T = D/c'$. Show that there is then a reference frame moving with speed V less than c in which the effect precedes the cause.

(a) Use Equ. 39-12. $t_2' - t_1' = \gamma[(t_2 - t_1) - (V/c^2)(x_2 - x_1)] = \gamma(T - VD/c^2)$, where $T = t_2 - t_1$ and $D = x_2 - x_1$.

(b) Events 1 and 2 are simultaneous in S' if $t_2' = t_1'$ or $(T - VD/c^2) = 0$. Since $V \le c,\ D \ge cT$.

(c) If $D < cT$ then $t_2' > t_1'$ and the events are not simultaneous in S'.

(c) If $D = c'T > cT$ then $T - VD/c^2 = T[1 - (V/c)(c'/c)] = t_2' - t_1'$. In this case, $t_2' - t_1'$ could be negative, i.e., t_2' could be less than t_1', or the effect could precede the cause.

Problems 25 through 29 refer to the following situation: An observer in S' lays out a distance $L' = 100$ light-minutes between points A' and B' and places a flashbulb at the midpoint C'. She arranges for the bulb to flash and for clocks at A' and B' to be started at zero when the light from the flash reaches them (see Figure 39-13). Frame S' is moving to the right with speed $0.6c$ relative to an observer C in S who is at the midpoint between A' and B' when the bulb flashes. At the instant he sees the flash, observer C sets his clock to zero.

25* ••   What is the separation distance between clocks A' and B' according to the observer in S?

L' is the proper distance L_p; use Equ. 39-14 $\gamma = 1.25;\ L = 80\ c\cdot$min

---

29*••   The time interval calculated in Problem 28 is the amount that the clock at A' leads that at B' according to the observer in S. Compare this result with L_pV/c^2.

The time interval calculated in Problem 28 is 60 min $L_pV/c^2 = 100\times0.6$ min $= 60$ min, as expected

---

33* •   A distant galaxy is moving away from us at a speed of 1.85×10^7 m/s. Calculate the fractional redshift $(\lambda' - \lambda_0)/\lambda_0$ in the light from this galaxy.

$(\lambda - \lambda_0)/\lambda_0 = \lambda/\lambda_0 - 1 = \sqrt{\dfrac{1 + V/c}{1 - V/c}} - 1$ $V/c = 0.185/3 = 0.0617;\ \lambda/\lambda_0 - 1 = 0.0637$

37* •   Herb and Randy are twin jazz musicians who perform as a trombone–saxophone duo. At the age of twenty, however, Randy got an irresistible offer to join a road trip to perform on a star 15 light-years away. To celebrate his bounteous luck, he bought a new vehicle for the trip—a deluxe space-coupe which could do $0.999c$. Each of the twins promises to practice diligently, so they can re-unite afterwards. Randy's gig goes so fabulously well, however, that he stays for a full 10 years before returning to Herb. After their reunion, (a) how many years of practice will Randy have? (b) how many years of practice will Herb have?

(b) 1. Find Δt_{travel} in Herb's frame, Δt_{travel} (H) Δt_{travel} (H) = (30 c·y)/(0.999c) = 30.03 y

2. Total time for Herb is 10 y + Δt_{travel} (H)   Herb's years of practice = 40.04 y

(a) 1. Find Δt_{travel} in Randy's frame, Δt_{travel} (R) γ = 22.37; Δt_{travel} (R) = (30.03 y)/22.37 = 1.34 y

2. Total time for Randy is Δt_{travel} (R) + 10 y   Randy's years of practice = 11.34 y

---

41* •   A spaceship is moving east at speed $0.90c$ relative to the earth. A second spaceship is moving west at speed $0.90c$ relative to the earth. What is the speed of one spaceship relative to the other?

Let S be the earth reference frame and S' be that of the ship traveling east (positive x direction). Then in the reference frame S', the velocity of S is directed west, i.e., $V = -u_x$. Now apply the velocity transformation equation, Equ. 39-19a to determine the speed of the other ship in the reference frame S'.

$$u_x' = \frac{u_x - V}{1 - Vu_x/c^2} = \frac{2u_x}{1 + u_x^2/c^2}$$ $u_x = -0.9c$; $u_x' = 0.994c$

45* •   Find the ratio of the total energy to the rest energy of a particle of rest mass m_0 moving with speed (a) $0.1c$, (b) $0.5c$, (c) $0.8c$, and (d) $0.99c$.

(a), (b), (c), (d) From Equ. 39-25, $E/E_0 = \gamma$ (a) $\gamma = 1.005$; (b) $\gamma = 1.15$; (c) $\gamma = 1.67$; (d) $\gamma = 7.09$

49* •   What is the energy of a proton whose momentum is $3m_0c$?

Use $E = c\sqrt{p^2 + (m_0c)^2}$ (see Problem 52) $E = \sqrt{10}\ m_0c^2 = 2.97$ GeV

53* ••   Use the binomial expansion (Equation 39-27) and Equation 39-28 to show that when $pc \ll m_0c^2$, the total energy is given approximately by

$$E \approx m_0c^2 + \frac{p^2}{2m_0}$$

From Equ. 39-28 we have $E = \sqrt{p^2c^2 + (E_0)^2} = m_0c^2\sqrt{1 + p^2/m_0^2c^2}$. When $p/m_0c \ll 1$ we can expand the square root, retaining only the first two terms. Thus, $E = m_0c^2[1 + \frac{1}{2}(p^2/m_0^2c^2)] = m_0c^2 + p^2/2m_0$; Q.E.D.

57* ••   The K^0 particle has a rest mass of 497.7 MeV/c^2. It decays into a π^- and π^+, each with rest mass 139.6 MeV/c^2. Following the decay of a K^0, one of the pions is at rest in the laboratory. Determine the kinetic energy of the other pion and of the K^0 prior to the decay.

We shall first consider the decay process in the center of mass reference frame and then transform to the laboratory reference frame in which one of the pions is at rest.

1. Write the conditions for energy conservation in CM $m_{K_0}c^2 = 2m_{\pi_0}\gamma c^2$; $\gamma = m_{K_0}/2m_{\pi_0} = 1.78$

2. Since one of the pions is at rest in the lab frame,

$\gamma = 1.78$ for the transformation to the lab frame;

find K of K^0 $K_K = 0.78 \times 497.7$ MeV = 389.5 MeV

3. Find total initial energy in lab frame $E = (497.7 + 389.5)$ MeV = 887.2 MeV

4. $K_\pi = E - 2m_{0\pi}c^2$ $K_\pi = (887 - 2 \times 139.6)$ MeV = 608 MeV

61* •••   A particle of rest mass 1 MeV/c^2 and kinetic energy 2 MeV collides with a stationary particle of rest mass 2 MeV/c^2. After the collision, the particles stick together. Find (a) the speed of the first particle before the collision, (b) the total energy of the first particle before the collision, (c) the initial total momentum of the system, (d) the total kinetic energy after the collision, and (e) the rest mass of the system after the collision.

(a) $E = K + E_0 = \gamma E_0$; $u/c = \sqrt{1 - 1/\gamma^2}$ $\gamma = 3$; $u = c\sqrt{8/9} = 0.943c$

(b) $E = \gamma E_0$ $E = 3E_0 = 3$ MeV

(c) $p^2c^2 = E^2 - E_0^2$; $p = \sqrt{E^2 - E_0^2}/c$ $p = \sqrt{8}E_0/c = 2.828$ MeV/c

(d), (e) 1. From energy conservation, $E_f = E_i$ $E_f = 5$ MeV

$\quad\quad$ 2. $p_f = p_i$; $E_f^2 = p_f^2c^2 + E_{f0}^2$ $E_{f0} = 4.123$ MeV $= m_{0f}c^2$; $m_{0f} = 4.123$ MeV/c^2

$\quad\quad$ 3. $K_f = E_f - E_{f0}$ $K_f = 0.877$ MeV

65* •   An observer sees a system consisting of a mass oscillating on the end of a spring moving past at a speed u and notes that the period of the system is T. Another observer, who is moving with the mass–spring system, also measures its period. The second observer will find a period that is (a) equal to T, (b) less than T, (c) greater than T, (d) either (a) or (b) depending on whether the system was approaching or receding from the first observer, (e) There is not sufficient information to answer the question.

(b)

69* •   How fast must a muon travel so that its mean lifetime is 46 μs if its mean lifetime at rest is 2 μs?

Find γ from Equ. 39-13; then $u/c = \sqrt{1 - 1/\gamma^2}$ $\gamma = 23$; $u = 0.999c$

73* ••   If a plane flies at a speed of 2000 km/h, for how long must it fly before its clock loses 1 s because of time dilation?

$\Delta t - \Delta t_p = \Delta t(1 - 1/\gamma) \approx \Delta t V^2/2c^2$ (see Problem 14) $\Delta t = [(2 \times 9 \times 10^{16})/(555.5)^2]$ s $= 5.83 \times 10^{11}$ s $\approx 1.85 \times 10^4$ y

---

77* ••   An interstellar spaceship travels from the earth to a distant star system 12 light-years away (as measured in the earth's frame). The trip takes 15 years as measured on the ship. (a) What is the speed of the ship relative to the earth? (b) When the ship arrives, it sends a signal to the earth. How long after the ship leaves the earth will it be before the earth receives the signal?

(a) 1. $\Delta t' = L'/u = L/\gamma u$ $\gamma u = (12 \; c \cdot y)/(15 \; y) = 0.8c$

$\quad\quad$ 2. Solve for u/c; $u/c = \sqrt{\dfrac{(\gamma u/c)^2}{1 + (\gamma u/c)^2}}$ $u/c = \sqrt{0.64/1.64} = 0.625$; $u = 0.625c$

(b) $T = L/u + L/c$ $T = (12/0.625 + 12)$ y $= 31.2$ y

81* •••   A rocket with a proper length of 700 m is moving to the right at a speed of 0.9c. It has two clocks, one in the nose and one in the tail, that have been synchronized in the frame of the rocket. A clock on the ground and the nose clock on the rocket both read $t = 0$ as they pass. (a) At $t = 0$, what does the tail clock on the rocket read as seen by an observer on the ground? When the tail clock on the rocket passes the ground clock, (b) what does the tail clock read as seen by an observer on the ground, (c) what does the nose clock read as seen by an observer on the ground, and (d) what does the nose clock read as seen by an observer on the rocket? (e) At $t = 1$ h, as measured on the rocket, a light signal is sent from the nose of the rocket to an observer standing by the ground clock. What does the ground clock read when the observer receives this signal? (f) When the observer on the ground receives the signal, he sends a return signal to the nose of the rocket. When is this signal received at the nose of the rocket as seen on the rocket?

We shall use the following notation: S is the ground reference frame, S' is the reference frame of the rocket, and $V = 0.9c$ is the speed of the rocket relative to S. We denote by T and N the tail and nose of the rocket, respectively.

(a) 1. Write the initial conditions in S' $t_N' = 0,\ x_N' = 0,$ and $t_T' = 0,\ x_T' = -L' = -700$ m

    2. Find x_T using length contraction $x_T = -L'/\gamma$

    3. Find t_T' at $t = 0$ using Equ. 39-12 $t_T' = \gamma(-Vx_T/c^2) = VL'/c^2 = 0.9 \times 700/c = 2.1\ \mu s$

(b) Find the time for rocket to move a distance L' $t_T' = L'/V = 700/0.9 \times 3 \times 10^8\ s = 2.59\ \mu s$

(c) $t_N = \Delta t' = (2.59 - 2.1)\ \mu s$ $t_N = 0.49\ \mu s$

(d) $t_N' = t_T'$ (clocks are synchronized in S') $t_N' = 2.59\ \mu s$

(e) 1. Find Δt, time is sent; use Equ. 39-13 $\Delta t = 2.294 \times 1$ h $= 2.294$ h

    2. Find Δt_{travel}, time of travel of signal to ground $\Delta t_{travel} = \Delta x/c = 2.294 \times 0.9$ h $= 2.065$ h

    3. Find t_{rec}, time the signal is received on ground $t_{rec} = (2.294 + 2.065)$ h $= 4.36$ h

(f) 1. Find Δx when signal is sent $\Delta x = (4.36\ h)(0.9c) = 3.924\ c \cdot h$

    2. In S, signal travels at $0.1c$ relative to rocket; find time t when signal reaches rocket. $\Delta t = (4.36\ h)(0.9c/0.1c) = 39.24$ h

         $t = (39.24 + 3.924)$ h $= 43.16$ h

    3. Use Equ. 39-13 to find t_N' $t_N' = (43.16/2.294)$ h $= 18.8$ h

---

85* ••• When a projectile particle with kinetic energy greater than the threshold kinetic energy K_{th} strikes a stationary target particle, one or more particles may be created in the inelastic collision. Show that the threshold kinetic energy of the projectile is given by

$$K_{th} = \frac{(\Sigma m_{in} + \Sigma m_{fin})(\Sigma m_{fin} - \Sigma m_{in})c^2}{2 m_{target}}$$

Here Σm_{in} is the sum of the rest masses of the projectile and target particles, Σm_{fin} is the sum of the rest masses of the final particles, and m_{target} is the rest mass of the target particle. Use this expression to determine the threshold kinetic energy of protons incident on a stationary proton target for the production of a proton–antiproton pair; compare your result with that of Problem 60.

In solving this problem we shall adopt the convention of the problem statement and use m to denote *rest masses* rather than relativistic masses. Let m_i denote the mass of the incident (projectile) particle. Then $\Sigma m_{in} = m_i + m_{target}$. Consider now the situation in the center of mass reference frame. At threshold we have $E^2 - p^2c^2 = \Sigma m_{fin}c^2$. Note that this is a relativistically invariant expression. In the laboratory frame, the target is at rest so $E_{target} = E_t = E_{t,0}$. We can therefore write $(E_i + E_{t,0})^2 - p_i^2c^2 = (\Sigma m_{fin}c^2)^2$. For the incident particle, $E_i^2 - p_i^2c^2 = E_{i,0}^2$ and $E_i = E_{i,0} + K_{th}$, where K_{th} is the threshold kinetic energy of the incident particle in the laboratory frame. We can now express K_{th} in terms of the rest energies: $(E_{t,0} + E_{i,0})^2 + 2K_{th}E_{t,0} = (\Sigma m_{fin}c^2)^2$. But $E_{t,0} + E_{i,0} = \Sigma m_{in}c^2$ and $E_{t,0} = m_{target}c^2$. Solving for K_{th} we obtain

$$K_{th} = \frac{(\Sigma m_{in} + \Sigma m_{fin})(\Sigma m_{fin} - \Sigma m_{in})c^2}{2 m_{target}}$$

For the creation of a proton - antiproton pair in a proton - proton collision, $\Sigma m_{in} = 2m_p$, $\Sigma m_{fin} = 4m_p$, and $m_{target} = m_p$. The above expression then gives $K_{th} = (6 \times 2/2)m_p c^2 = 6m_p c^2$, where here m_p denotes the rest mass of a proton.

89* ••• For the special case of a particle moving with speed u along the y axis in frame S, show that its momentum and energy in frame S' are related to its momentum and energy in S by the transformation equations

$$P_x' = \gamma\left(p_x - \frac{VE}{c^2}\right), \quad p_y' = p_y, \quad p_z' = p_z'; \qquad \frac{E'}{c} = \gamma\left(\frac{E}{c} - \frac{Vp_x}{c^2}\right)$$

Compare these equations with the Lorentz transformation for x', y', z', and t'. These equations show that the quantities p_x, p_y, p_z, and E/c transform in the same way as do x, y, z, and ct.

In S, $u_x = u_z = 0$, $u_y = u$; $p_x = p_z = 0$, $p_y = \gamma_u m_0 u$, and $E = \gamma_u m_0 c^2$. Here, $\gamma_u = 1/\sqrt{1 - u^2/c^2}$. Then, applying the velocity transformation equations we find, in S', $u_x' = -V$, $u_y' = u/\gamma$, $u_z' = 0$. This gives $u'^2 = V^2 + u^2(1 - V^2/c^2) = V^2 + u^2 - V^2u^2/c^2$ and $(1 - u'^2/c^2) = 1 - V^2/c^2 - u^2/c^2 + V^2u^2/c^4 = (1 - V^2/c^2)(1 - u^2/c^2)$. In S' the momentum components are $p_x' = \gamma' m_0 u_x'$, $p_y' = \gamma' m_0 u_y'$, and $p_z' = 0$, where $\gamma' = 1/\sqrt{1 - u'^2/c^2} = \gamma_u/\sqrt{1 - V^2/c^2}$. In terms of the parameters in S, $p_x' = -\gamma_u m_0 V/\sqrt{1 - V^2/c^2} = -\gamma EV/c^2$, where $\gamma = 1/\sqrt{1 - V^2/c^2}$. Since $p_x = 0$, $p_x' = \gamma(p_x - EV/c^2)$. In terms of the parameters in S, $p_y' = p_y$ and $p_z' = p_z$. $E' = \gamma' m_0 c^2 = \gamma\gamma_u m_0 c^2 = \gamma E = \gamma(E - Vp_x/c)$ and $E'/c = \gamma(E/c - Vp_x/c^2)$. Note that $E'^2 - p'^2c^2 = E_0^2$ (see Problem 91), which demonstrates that E_0 is a relativistic invariant. Also, comparison with Equs. 39-11 and 39-12 shows that the components of p and E/c transform as do the components of r and ct.

CHAPTER 40

Nuclear Physics

1* • Give the symbols for two other isotopes of (a) ^{14}N, (b) ^{56}Fe, and (c) ^{118}Sn

(a) ^{15}N, ^{16}N; (b) ^{54}Fe, ^{55}Fe; (c) ^{114}Sn, ^{116}Sn

---

5* • (a) Given that the mass of a nucleus of mass number A is approximately $m = CA$, where C is a constant, find an expression for the nuclear density in terms of C and the constant R_0 in Equation 40-1. (b) Compute the value of this nuclear density in grams per cubic centimeter using the fact that C has the approximate value of 1 g per Avogadro's number of nucleons.

(a) From Equ. 40-1, $R = R_0 A^{1/3}$, the nuclear volume is $V = (4\pi/3)R_0^3 A$. With $m = CA$, $\rho = m/V = 3C/4\pi R_0^3$.

(b) Given that $C = 1/6.02 \times 10^{23}$ g and $R_0 = 1.5 \times 10^{-13}$ cm, $\rho = 1.18 \times 10^{14}$ g/cm^3.

9* •• (a) Calculate the radii of $^{141}_{56}$Ba and $^{92}_{36}$Kr from Equation 40-1. (b) Assume that after the fission of ^{235}U into ^{141}Ba and ^{92}Kr, the two nuclei are momentarily separated by a distance r equal to the sum of the radii found in (a), and calculate the electrostatic potential energy for these two nuclei at this separation. (See Problem 8.) Compare your result with the measured fission energy of 175 MeV.

(a) Use Equ. 40-1, with $A = 141$ and 92, respectively

(b) Use $ke^2 = 1.44$ MeV·fm; $U = kZ_1 Z_2 e^2/(r_1 + r_2)$

For $^{141}_{56}$Ba, $R = 7.81$ fm ; for $^{92}_{36}$Kr, $R = 6.77$ fm

$U = (1.44 \times 56 \times 36/14.58)$ MeV = 199 MeV; this is somewhat greater than the fission energy of 175 MeV

13* • What effect would a long-term variation in cosmic-ray activity have on the accuracy of ^{14}C dating?

It would make the dating unreliable because the current concentration of ^{14}C is not equal to that at some earlier time.

17* • The half-life of radium is 1620 y. Calculate the number of disintegrations per second of 1 g of radium, and show that the disintegration rate is approximately 1 Ci.

1. Use Equ. 40-11 to find λ

2. Determine $N_0 = N_A/m$

3. Find R from Equ. 40-7

$\lambda = (0.693/1620)$ y^{-1} = 4.28×10^{-4} y^{-1} = 1.35×10^{-11} s^{-1}

$N_0 = 6.02 \times 10^{23}/226 = 2.664 \times 10^{21}$

$R = (1.35 \times 10^{-11} \times 2.664 \times 10^{21})$ s^{-1} = 3.60×10^{10} s^{-1} ≈ 3.7×10^{10} s^{-1} = 1 Ci

---

21* • At the scene of the crime, in the museum's west wing, Angela found some wood chips, so she slipped them into her purse for future analysis. They were allegedly from an old wooden mask, which the guard said he threw at the would-be thief. Later, in the lab, she determined the age of the chips, using a sample which contained 10 g of carbon and showed a ^{14}C decay rate of 100 counts/min. How old are they?

1. Find $R/R_0 = (1/2)^n$ $R_0 = (15 \times 10) \text{ s}^{-1} = 150 \text{ s}^{-1}; (1/2)^n = 2/3$

2. Solve for $n = [\ln (R/R_0)]/[\ln (0.5)]$ $n = 0.585$

3. Find $t = n\tau_{1/2}$ $t = (0.585 \times 5730) \text{ y} = 3352 \text{ y}$

25* •• Derive the result that the activity of 1 g of natural carbon due to the β decay of ^{14}C is 15 decays/min = 0.25 Bq.

1. Find number of ^{14}C per gram of C $N(^{14}C) = (6.02 \times 10^{23}/12)(1.3 \times 10^{-12}) = 6.52 \times 10^{10}$

2. Find the decay constant $\lambda = (0.693/5730) \text{ y}^{-1} = 3.83 \times 10^{-12} \text{ s}^{-1} = 2.30 \times 10^{-10} \text{ min}^{-1}$

3. $R = \lambda N$ $R = (6.52 \times 10^{10} \times 2.30 \times 10^{-10}) \text{ min}^{-1} = 1.50 \text{ min}^{-1}$

29* •• A 1.00-mg sample of substance of atomic mass 59.934 u emits β particles with an activity of 1.131 Ci. Find the decay constant for this substance in s^{-1} and its half-life in years.

1. Find N $N = (6.022 \times 10^{23} \times 10^{-3}/59.934) = 1.005 \times 10^{19}$

2. $\lambda = R/N$ (see Equ. 40-7) $R = 1.131 \times 3.7 \times 10^{10}; \lambda = 4.165 \times 10^{-9} \text{ s}^{-1}$

3. Use Equ. 40-11 $\tau_{1/2} = (0.693/4.165 \times 10^{-9}) \text{ s} = 1.664 \times 10^8 \text{ s} = 5.27 \text{ y}$

33* ••• If there are N_0 radioactive nuclei at time $t = 0$, the number that decay in some time interval dt at time t is $-dN = \lambda N_0 e^{-\lambda t} dt$. If we multiply this number by the lifetime t of these nuclei, sum over all the possible lifetimes from $t = 0$ to $t = \infty$, and divide by the total number of nuclei, we get the mean lifetime τ.

$$\tau = \frac{1}{N_0}\int_0^\infty t|dN| = \int_0^\infty t\lambda e^{-\lambda t}$$

Show that $\tau = 1/\lambda$.

Note that λ is a constant; so $\int\limits_0^\infty t\lambda e^{-\lambda t} dt = \frac{1}{\lambda}\int\limits_0^\infty x e^{-x} dx$. The definite integral has the value 1, so $\tau = 1/\lambda$.

37* •• (a) Use the atomic masses $m = 13.00574$ u for $^{13}_{7}N$ and $m = 13.003354$ u for $^{13}_{6}C$ to calculate the Q value (in MeV) for the β decay

$$^{13}_{7}N \rightarrow {}^{13}_{6}C + \beta^+ + \nu_e$$

(b) Explain why you need to add two electron masses to the mass of $^{13}_{6}C$ in the calculation of the Q value for this reaction.

(a) for β^+ decay, $Q = (m_i - m_f - 2m_e)c^2$ $Q = (0.002386 \times 931.5 - 2 \times 0.511) \text{ MeV} = 1.20 \text{ MeV}$

(b) The atomic masses include the masses of the electrons of the neutral atoms. In this reaction the initial atom has 7 electrons, the final atom only has 6 electrons. Moreover, in addition to the one electron not included in the atomic masses, a positron of mass equal to that of an electron is created. Consequently, one must add the rest energies of two electrons to the rest energy of the daughter atomic mass when calculating Q.

41* • What happens to the neutrons produced in fission that do not produce another fission?

Some of the neutrons are captured by other nuclei. Those not captured decay according to the reaction

$$n \rightarrow p + e + \bar{\nu}_e.$$

45* • Compute the temperature T for which $kT = 10$ keV, where k is Boltzmann's constant.

$k = 8.62 \times 10^{-5}$ eV/K $= 8.62 \times 10^{-8}$ keV/K $T = 10/8.62 \times 10^{-8} = 1.16 \times 10^8 \text{ K}$

49* ••• Energy is generated in the sun and other stars by fusion. One of the fusion cycles, the proton–proton cycle, consists of the following reactions:

$$^1\text{H} + {}^1\text{H} \rightarrow {}^2\text{H} + \beta^+ + \nu_e$$
$$^1\text{H} + {}^2\text{H} \rightarrow {}^3\text{He} + \gamma$$

followed by

$$^1\text{H} + {}^3\text{He} \rightarrow {}^4\text{He} + \beta^+ + \nu_e$$

(a) Show that the net effect of these reactions is

$$4\,{}^1\text{H} \rightarrow {}^4\text{He} + 2\beta^+ + 2\nu_e + \gamma$$

(b) Show that rest energy of 24.7 MeV is released in this cycle (not counting the energy of 1.02 MeV released when each positron meets an electron and the two annihilate). (c) The sun radiates energy at the rate of about 4×10^{26} W. Assuming this is due to the conversion of four protons into helium plus γ rays and neutrinos, which releases 26.7 MeV, what is the rate of proton consumption in the sun? How long will the sun last if it continues to radiate at its present level? (Assume that protons constitute about half of the total mass $[2 \times 10^{30}$ kg] of the sun.)

(a) 1. Sum the three reactions $4\,{}^1\text{H} + {}^2\text{H} + {}^3\text{He} \rightarrow {}^2\text{H} + {}^3\text{He} + {}^4\text{He} + 2\beta^+ + 2\nu_e + \gamma$

   2. Cancel the common quantities $4\,{}^1\text{H} \rightarrow {}^4\text{He} + 2\beta^+ + 2\nu_e + \gamma$

(b) $\Delta mc^2 = (4m_p - m_\alpha - 4m_e)c^2$ (see Problem 37) $\Delta mc^2 = [(4 \times 1.007825 - 4.002603)931.5 - 4 \times 0.511]$ MeV
    $= 25.69$ MeV

(c) 1. Find N, the number of protons in the sun $N = 10^{30}/1.673 \times 10^{-27} = 5.98 \times 10^{56}$

   2. Find the energy released per proton in fusion $E = 26.7/4$ MeV $= 6.675$ MeV $= 1.07 \times 10^{-12}$ J

   3. Find R, the rate of proton consumption $R = P/E = 4 \times 10^{26}/1.07 \times 10^{-12}$ s$^{-1} = 3.745 \times 10^{38}$ s^{-1}

   4. Find T, the time for consumption of all protons $T = 5.98 \times 10^{56}/3.745 \times 10^{38}$ s $= 1.60 \times 10^{18}$ s $= 5.05 \times 10^{10}$ y

---

53* • Why does fusion occur spontaneously in the sun but not on earth?

Fusion requires extremely high temperature and pressure. These conditions are met in the core of the sun but not on earth.

57* • The isotope ^{14}C decays according to $^{14}\text{C} \rightarrow {}^{14}\text{N} + e^- + \overline{\nu}_e$. The atomic mass of ^{14}N is 14.003074 u. Determine the maximum kinetic energy of the electron. (Neglect recoil of the nitrogen atom.)

$E_{max} = Q = [m(^{14}\text{C}) - m(^{14}\text{N})]c^2$ $E_{max} = (0.003242 - 0.003074) \times 931.5$ MeV $= 156.5$ keV

61* • The relative abundance of ^{40}K (molecular mass 40.0 g/mol) is 1.2×10^{-4}. The isotope ^{40}K is radioactive with a half-life of 1.3×10^9 y. Potassium is an essential element of every living cell. In the human body the mass of potassium constitutes approximately 0.36% of the total mass. Determine the activity of this radioactive source in a student whose mass is 60 kg.

1. Find N, the number of K nuclei in the person $N = 60 \times 0.0036 \times 6.02 \times 10^{26}/39.1 = 3.326 \times 10^{24}$

2. Find N_{40}, the number of ^{40}K nuclei $N_{40} = 1.2 \times 10^{-4} \times 3.326 \times 10^{24} = 3.99 \times 10^{20}$

3. Find $N_{40}\lambda$, the activity of the ^{40}K nuclei $\lambda = [0.693/(1.3 \times 10^9 \times 3.156 \times 10^7)]$ s$^{-1} = 1.69 \times 10^{-17}$ s^{-1};
    activity $= 6.74 \times 10^3$ Bq

65* •• Twelve nucleons are in a one-dimensional infinite square well of length $L = 3$ fm. (a) Using the approximation that the mass of a nucleon is 1 u, find the lowest energy of a nucleon in the well. Express your answer in MeV.

What is the ground-state energy of the system of 12 nucleons in the well if (b) all the nucleons are neutrons so that there can be only 2 in each state and (c) 6 of the nucleons are neutrons and 6 are protons so that there can be 4 nucleons in each state? (Neglect the energy of Coulomb repulsion of the protons.)

(a) Use Equ. 36-13 with $n = 1$

$$E_1 = \frac{(6.626\times10^{-34})^2}{8 \times 1.66\times10^{-27} \times 9\times10^{-30}} \text{ J} = 22.96 \text{ MeV}$$

(b) Neutrons are fermions; only 2 per state;
$E_n = n^2 E_1$

$E = 2(E_1 + E_2 + E_3 + E_4 + E_5 + E_6) = 2\times91\times22.96$ MeV = 4.178 GeV

(c) Find E for 4 protons and 4 neutrons

$E = 4(E_1 + E_2 + E_3) = 56E_1 = 1.286$ GeV

69* •• (a) Find the wavelength of a particle in the ground state of a one-dimensional infinite square well of length $L = 2$ fm. (b) Find the momentum in units of MeV/c for a particle with this wavelength. (c) Show that the total energy of an electron with this wavelength is approximately $E \approx pc$. (d) What is the kinetic energy of an electron in the ground state of this well? This calculation shows that if an electron were confined in a region of space as small as a nucleus, it would have a very large kinetic energy.

(a) In ground state, $\lambda = 2L$ (see Equ. 17-17)

$\lambda = 4$ fm

(b) Use Equ. 17-7, $p = h/\lambda = hc/\lambda c$

$p = (1240 \text{ eV·nm})/[(4\times10^{-6} \text{ nm})c] = 310$ MeV/c

(c) $E^2 = E_0^2 + p^2c^2$; $E_0 = 0.511$ MeV $<< 310$ MeV

$E^2 \approx p^2c^2$; $E \approx pc$

(d) $K = E - E_0 \approx E$

$K \approx 310$ MeV

73* ••• (a) Use the result of part (e) of Problem 72 (Equation 40-23) to show that after N head-on collisions of a neutron with carbon nuclei at rest, the energy of the neutron is approximately $(0.714)^N E_0$, where E_0 is its original energy. (b) How many head-on collisions are required to reduce the energy of the neutron from 2 MeV to 0.02 eV, assuming stationary carbon nuclei?

(a) 1. Evaluate Equ. 40-23 for $m/M = 12$

$m = 1.00866u$; $\Delta E/E = -0.286$

2. Determine E_f/E_0 per collision

$E_f = E_0 + \Delta E = 0.714E_0$; $E_f/E_0 = 0.714$

3. For N collsions, $E_f/E_0 = 0.714^N$

$E_f = (0.714)^N E_0$

(b) $N = [\ln(E_f/E_0)]/[\ln(0.714)]$; $E_f/E_0 = 10^{-8}$

$N = 54.7$; number of collisions = 55

77* ••• An example of the situation discussed in Problem 75 is the radioactive isotope ^{229}Th, an α emitter with a half-life of 7300 years. Its daughter, ^{225}Ra, is a β emitter with a half-life of 14.8 d. In this, as in many instances, the half-life of the parent is much longer than that of the daughter. Using the expression given in Problem 75 (b), show that, starting with a sample of pure ^{229}Th containing N_{A0} nuclei, the number, N_B, of ^{225}Ra nuclei will, after a several years, be a constant, given by

$$N_B = (\lambda_A/\lambda_B)N_A$$

The number of daughter nuclei are said to be in "secular equilibrium."

Since $\tau_A >> \tau_B$, $\lambda_A << \lambda_B$. Then after a time interval of some ten τ_B, $(e^{-\lambda_A t} - e^{-\lambda_B t}) \approx 1$ since $\lambda_A t << 1$. The ratio $\lambda_A/(\lambda_B - \lambda_A) \approx \lambda_A/\lambda_B$ when $\lambda_A << \lambda_B$. Thus, the result of Problem 75(b) reduces to $N_B = (\lambda_A/\lambda_B)N_{A0}$. However, in the time interval of about ten τ_B, N_A has not diminished significantly, so $N_A \approx N_{A0}$ and $N_B = (\lambda_A/\lambda_B)N_A$.

CHAPTER 41

Elementary Particles and the Beginning of the Universe

1* • How are baryons and mesons similar? How are they different?

Similarities: Baryons and mesons decay via the strong interaction. Both are composed of quarks.

Differences: Baryons consist of three quarks and are fermions. Mesons consist of two quarks and are bosons.

Baryons have baryon number +1 or −1. Mesons have baryon number 0.

5* • True or false:

Mesons are spin-½ particles.

False

9* • Determine the change in strangeness in each reaction that follows, and state whether the reaction can proceed via the strong interaction, the weak interaction, or not at all: (a) $\Omega^- \rightarrow \Xi^0 + \pi^-$, (b) $\Xi^0 \rightarrow p + \pi^- + \pi^0$, and

(c) $\Lambda^0 \rightarrow p^+ + \pi^-$.

(a) 1. List S of Ω^-, Ξ^0, and π^-; see Fig. 41-2 Ω^-, $S = -3$; Ξ^0, $S = -2$; π^-, $S = 0$

     2. Determine ΔS $\Delta S = +1$; reaction via the weak interaction allowed

(b) 1. List S of Ξ^0, p, π^-, and π^0 Ξ^0, $S = -2$; p, $S = 0$; π^-, $S = 0$; π^0, $S = 0$

     2. Determine ΔS $\Delta S = +2$; reaction is not allowed

(c) 1. List S of Λ^0, p^+, and π^- Λ^0, $S = -1$; for p and π^-, $S = 0$

     2. Determine ΔS $\Delta S = +1$; reaction via the weak interaction allowed

---

13* •• Consider the following decay chain:

$\Omega^- \rightarrow \Xi^0 + \pi^-$

$\Xi^0 \rightarrow \Sigma^+ + e^- + \overline{\nu}_e$

$\pi^- \rightarrow \mu^- + \overline{\nu}_\mu$

$\Sigma^+ \rightarrow n + \pi^+$

$\pi^+ \rightarrow \mu^+ + \nu_\mu$

$\mu^+ \rightarrow e^+ + \overline{\nu}_\mu + \nu_e$

$\mu^- \rightarrow e^- + \overline{\nu}_e + \nu_\mu$

(a) Are all the final products shown stable? If not, finish the decay chain. (b) Write the overall decay reaction for Ω^- to the final products. (c) Check the overall decay reaction for the conservation of electric charge, baryon number, lepton number, and strangeness.

(a) No; the neutron is not stable: $n \rightarrow p^+ + e^- + \overline{v}_e$

(b) Adding the reactions one obtains $\Omega^- \rightarrow p^+ + e^+ + 3e^- + v_e + 3\overline{v}_e + 2\overline{v}_\mu + 2v_\mu$

(c) 1. Charge conservation: $-1 \rightarrow 1 + 1 - 3 = -1$; charge is conserved.

   2. Baryon number: $1 \rightarrow 1$; baryon number is conserved.

   3. Lepton number; electrons: $0 \rightarrow -1 + 3 + 1 - 3 = 0$; lepton number for electrons is conserved.

   Lepton number; muons: $0 \rightarrow -2 + 2 = 0$; lepton number for muons is conserved.

   4. Strangeness: $-3 \rightarrow 0$; strangeness is not conserved. However, in each baryon decay $\Delta S = +1$, and each decay is allowed via the weak interaction.

17* • Find the baryon number, charge, and strangeness for the following quark combinations and identify the hadron: (a) uud, (b) udd, (c) uus, (d) dds, (e) uss, and (f) dss.

(a) uud: Find B, q, and S; see Table 41-2 $B = 1$, $q = +e$, $S = 0$; hadron is p^+

(b) udd: Find B, q, and S; see Table 41-2 $B = 1$, $q = 0$, $S = 0$; hadron is n

(c) uus: Find B, q, and S; see Table 41-2 $B = 1$, $q = +e$, $S = -1$; hadron is Σ^+

(d) dds: Find B, q, and S; see Table 41-2 $B = 1$, $q = -e$, $S = -1$; hadron is Σ^-

(e) uss: Find B, q, and S; see Table 41-2 $B = 1$, $q = 0$, $S = -2$; hadron is Ξ^0

(f) dss: Find B, q, and S; see Table 41-2 $B = 1$, $q = -1$, $S = -2$; hadron is Ξ^-

---

21* • The D^+ meson has no strangeness, but it has charm of $+1$. (a) What is a possible quark combination that will give the correct properties for this particle? (b) Repeat (a) for the D^- meson, which is the antiparticle of the D^+.

(a) $B = 0$, so we must look for a combination of quark and antiquark. Since it has charm of $+1$, one of the quarks must be c. So that the charge is $+e$, the antiquark must be $\overline{d}$. The possible combination for D^+ is $c\overline{d}$.

(b) Since D^- is the antiparticle of D^+, the quark combination is $\overline{c}d$.

25* •• Find a possible quark combination for the following particles: (a) Ω^- and (b) Ξ^-.

(a) For Ω^-; $B = +1$, $q = -e$, $S = -3$. The quark combination that meets these conditions is sss.

(b) For Ξ^-; $B = +1$, $q = -e$, $S = -2$. The quark combination that meets these conditions is ssd.

29* •• Consider the following decay chain:

$\Xi^0 \rightarrow \Lambda^0 + \pi^0$

$\Lambda^0 \rightarrow p + \pi^-$

$\pi^0 \rightarrow \gamma + \gamma$

$\pi^- \rightarrow \mu^- + \overline{v}_\mu$

$\mu^- \rightarrow e^- + \overline{v}_e + v_\mu$

(a) Are all the final products shown stable? If not, finish the decay chain. (b) Write the overall decay reaction for Ξ^0 to the final products. (c) Check the overall decay reaction for the conservation of electric charge, baryon number, lepton number, and strangeness. (d) In the first step of the chain, could the Λ^0 have been a Σ^0?

(a) Yes, the final products are stable.

(b) The end result is $\Xi^0 \rightarrow p^+ + e^- + \overline{v}_e + v_\mu + \overline{v}_\mu + 2\gamma$.

(c) 1. Charge conservation: $0 \rightarrow e^+ + e^- = 0$; charge is conserved

   2. Baryon number: $1 \rightarrow 1 + 0 = 1$; baryon number is conserved.

   3. Strangeness: $-2 \rightarrow 0$; $\Delta S = +2$; the reaction is allowed via the weak interaction because in the first two decays

$\Delta S = +1$.

(d) No; the rest masses of the decay products would be greater than the rest mass of the Ξ^0, violating energy conservation.

33* ••• In this problem, you will calculate the difference in the time of arrival of two neutrinos of different energy from a supernova that is 170,000 light-years away. Let the energies of the neutrinos be $E_1 = 20$ MeV and $E_2 = 5$ MeV, and assume that the rest mass of a neutrino is 20 eV/c^2. Because their total energy is so much greater than their rest energy, the neutrinos have speeds that are very nearly equal to c and energies that are approximately $E \approx pc$. (a) If t_1 and t_2 are the times it takes for neutrinos of speeds u_1 and u_2 to travel a distance x, show that

$$\Delta t = t_1 - t_2 = x\frac{u_1 - u_2}{u_1 u_2} \approx \frac{x\,\Delta u}{c^2}$$

(b) The speed of a neutrino of rest mass m_0 and total energy E can be found from Equation 39-25. Show that when $E \gg m_0 c^2$, the speed u is given approximately by

$$\frac{u}{c} \approx 1 - \frac{1}{2}\left(\frac{m_0 c^2}{E}\right)^2$$

(c) Use the results for (b) to calculate $u_1 - u_2$ for the energies and rest mass given, and calculate Δt from the result for (a) for $x = 170{,}000 c\cdot$y. (d) Repeat the calculation in (c) using $m_0 c^2 = 40$ eV for the rest energy of a neutrino.

(a) 1. Express $\Delta t = t_2 - t_1$ in terms of u_1 and u_2

$$\Delta t = \frac{x}{u_2} - \frac{x}{u_1} = \frac{x(u_1 - u_2)}{u_1 u_2}$$

 2. Note that $u_1 u_2 \approx c^2$; let $\Delta u = u_1 - u_2$

$$\Delta t = \frac{x\,\Delta u}{c^2}$$

(b) 1. Use Equ. 39-25 to write u/c

$$u/c = \sqrt{1 - (m_0 c^2/E)^2}$$

 2. Use the binomial expansion; $(m_0 c^2/E)^2 \ll 1$

$$u/c \approx 1 - \tfrac{1}{2}(m_0 c^2/E)^2$$

(c) 1. Write $u_1 - u_2$ in terms of E_1 and E_2 and $m_0 c^2$

$$u_1 - u_2 = \tfrac{1}{2}c(m_0 c^2)^2\left(\frac{1}{E_2^2} - \frac{1}{E_1^2}\right) = \frac{c(m_0 c^2)^2(E_1^2 - E_2^2)}{2 E_1^2 E_2^2}$$

 2. Evaluate Δt for $m_0 c^2 = 20$ eV, $E_1 = 20$ MeV, $E_2 = 5$ MeV, and $x = 170{,}000\ c\cdot$y

$\Delta u = 7.5\times10^{-12} c$; $\Delta t = 1.275\times10^{-6}$ y $= 40.2$ s

(d) Repeat (c) for $m_0 c^2 = 40$ eV

$\Delta u = 4\times 7.5\times10^{-12} c = 3\times10^{-11} c$; $\Delta t = 4\times 40.2$ s $= 161$ s

Note that the spread in the arrival time for neutrinos from a supernova can be used to estimate the rest mass of a neutrino.